SHRIMPING
WEST TEXAS

SHRIMPING WEST TEXAS

THE RISE AND FALL OF THE PERMIAN SEA SHRIMP COMPANY

BART REID

TEXAS TECH UNIVERSITY PRESS

This book is typeset in EB Garamond. The paper used in this book meets the minimum requirements of ANSI/NISO Z39.48-1992 (R1997). ♾

Designed by Hannah Gaskamp
Cover design by Hannah Gaskamp

Library of Congress Cataloging-in-Publication Data

Names: Reid, Bart, 1963– author. Title: Shrimping West Texas: The Rise and Fall of the Permian Sea Shrimp Company / Bart Reid. Description: Lubbock, Texas: Texas Tech University Press, 2024. | Includes index. | Summary: "The story of a uniquely West Texas aquaculture enterprise: the Permian Sea Shrimp Company"—Provided by publisher.
Identifiers: LCCN 2023050785 (print) | LCCN 2023050786 (ebook) | ISBN 978-1-68283-211-0 (paperback) | ISBN 978-1-68283-212-7 (ebook)
Subjects: LCSH: Reid, Bart, 1963– | Shrimp culture—Permian Basin (Tex. and N.M.). | Shrimp culture—Texas.
Classification: LCC SH380.62.U6 R45 2024 (print) | LCC SH380.62.U6 (ebook) |
DDC 338.7/6396809764—dc23/eng/20240310
LC record available at https://lccn.loc.gov/2023050785
LC ebook record available at https://lccn.loc.gov/2023050786

24 25 26 27 28 29 30 31 32 / 9 8 7 6 5 4 3 2 1

Texas Tech University Press
Box 41037
Lubbock, Texas 79409-1037 USA
800.832.4042
ttup@ttu.edu
www.ttupress.org

To my wife Patsy and my children Kolton, Ryan, and Hannah
who lived it and maybe loved it, or parts of it anyway.
And to my Aunt Jeannette Brown, who gave me the first investment money for Permian Sea Shrimp Company and told me to
follow my dream.
Also dedicated to all the good folks of West Texas who loved the
shrimp and the farms as much as we did.

A PASSERBY ASKS THE AGGIE FARMER: *WHAT ARE YOU GROWING THERE?*

Aggie farmer: *This here is a shrimp farm.*

Passerby: *It don't look like you made much of a crop from what I can tell.*

Aggie farmer: *No sir, we didn't. I think next year we should plant 'em a little deeper in the ground.*

CONTENTS

CONTENTS

ILLUSTRATIONS

FOREWORD

In aquaculture there is a timeless, unwritten adage that states, "If something can go wrong, you can bet it will." Along with that maxim follows, "The ones that persevere (with a big dose of Luck) will have a decent chance of coming out victorious in the end." Both precepts mirror the blow-by-blow saga Bart Reid relives in his book about growing marine shrimp (native to the Pacific) in the West Texas desert. Bart has done a splendid job of describing the many ups and downs of a start-up shrimp farmer operating in a seemingly hostile environment. He faced all kinds of unexpected and near disasters and dodged most of them. There were countless obstacles to overcome, hurdles to clear, disappointments, good and bad ideas, setbacks, frustrations, and finally celebrations at harvest time. Turns out the water pumped up 200 feet from the ancient Permian Sea is of an ideal salinity, and Bart came out a winner along with the good folks in and around Imperial, Texas, who loved the product.

I met Bart in 1986 when he applied to the master of science in mariculture degree program at Texas A&M–Corpus Christi. I had developed this graduate program emphasizing both science and business, and it attracted students from around the world. By the time I retired in 2013, the program had accepted and graduated students from twenty-one US states and seventeen foreign countries: bright, driven, and industrious students who wanted to learn as much as they could about growing commercially important species of marine fish and shrimp. And the Corpus Christi area was the place to be because of the myriad hands-on opportunities within thirty minutes of campus for the students to experience.

All students applying to the program were strongly encouraged to visit campus for an interview and take tours of local culture facilities. Bart arrived with a "Put me in, Coach" enthusiasm. It was clear that he knew what he wanted and seemed to know how to get there. One option in the program was to conduct research, and that's the route he took. He was most fortunate to land a spot with Dr. Connie Arnold and Dr. Joan Holt in Port Aransas with the Fisheries and Mariculture Lab at the University of Texas Marine Science Institute. Their research involved the culture of select marine fish and shrimp in closed-system indoor raceways at much higher densities than were possible in outdoor ponds. This is where Bart got his start, and he had the good fortune of having these renowned researchers to oversee and guide his work.

Bart's research thesis focused on the growth and survival of the Pacific white shrimp stocked at extremely high densities where a complete diet must be supplied, and the animals must grow well with it. In these systems, pH, temperature, and salinity requirements of the animals must be met and maintained tightly 24/7. Biological filters must be maintained and effective at removing toxic ammonia by-products due to animal excretion, uneaten feed, etc. As the animals increase in size, they require more oxygen and more feed; they also produce more waste, and the chances of a disease outbreak increase exponentially. It is not uncommon to hear of catastrophic shrimp mortality near the end of the growing period. It's been said that you are not an aquaculturist until you've lost an entire crop of animals. If true, then I am one through and through. Restless nights filled with "what if" nightmares are the norm around harvest time. I believe Bart's exposure to "high risk culture in search of high production" during graduate school with Arnold and Holt instilled in him the "never say never" attitude that served him well in his farming career.

I should add that while Bart was finding success in the desert, shrimp farms along the Texas coast were having little success; during the 1990s most had ceased operations. Disease was a major culprit, and discharging pond water containing pathogens was a significant issue. The capture of escaped exotic shrimp in public waters was another. Cheaper imported shrimp kept farm prices down. On the flip side, Bart's groundwater quality was ideal, was present in an endless supply, and was free of pathogens, and water discharge was a nonissue. He had no problem selling all he could grow locally at a premium price. Any shrimp that escaped from his farm had no chance of showing up in a Texas bay.

Bart proved that successful commercial culture of shrimp in the desert wasn't such a crazy idea. I love a success story, and this is a good one!

DAVID A. MCKEE

PROFESSOR EMERITUS,

TEXAS A&M UNIVERSITY–CORPUS CHRISTI

PREFACE

The Trans-Pecos is the region of Texas that lies west of the Pecos River. It is bounded by the Rio Grande to the south and the state of New Mexico to the north. It is also located right smack in the Chihuahua Desert—a dry, desolate desert with little surface water and a harsh climate. Lots of folks passed through this desert over the course of history. The Spanish explorers Álvar Núñez Cabeza de Vaca, Antonio de Espejo, and Gaspar Castaño de Sosa all explored the Trans-Pecos in the 1500s and hightailed it out of there in short order. Even the Comanche who ruled all the Southwest from Kansas to Mexico for two hundred years passed through the Trans-Pecos on their winter migration to Mexico, only stopping to rest their horses and steel their resolve to keep moving to somewhere better. For six hundred years, this land was "pass through country," much like Middle America became "flyover country" to the coastal elites. From the cattle drives to the mail routes, these were just bad lands to get through as fast as possible. There were no real settlers here until about the 1880s. To stay and settle in the hardscrabble of the Trans-Pecos you either have to be out of options or have a vision for something remarkable. I had a vision, and I wanted to make it remarkable. For all the area's negative attributes, I could see there was great potential—potential for a thriving shrimp farming industry like no other.

West Texas is blessed with special resources including a dry climate, good infrastructure thanks to the oil field, and lots and lots of salty water. Typically, deserts have saline groundwater if they have groundwater at all, and the Trans-Pecos is no exception. It sits directly on top of the saltwater aquifer known as the Permian Sea. Even the Pecos River in this area runs as salty as most Texas bay systems. Salt water cannot grow crops or water livestock, so using it for aquaculture makes perfect sense. I wish I could say that I found and recognized this resource and established this whole idea and the inland aquaculture industry, but I cannot. A number of folks recognized this resource and all its potential and started the process years before I came along. As a matter of fact, when I was still a graduate student a wealthy oil man contacted me about running a redfish farm for him and his son in Midland. He had purchased some large fiberglass tanks and put some big redfish brood stock in them to start a hatchery to supply baby fish for his soon-to-be-built farm. Without any expert help, he knew that he and the son were in way over their heads. He flew my wife and me to Midland, and we toured the little would-be hatchery and future farm site. What is ironic was that I thought the venture too crazy to consider, just too far out there—something from science fiction.

Some five years later I would be singing a very different tune. I became a true believer, so I decided to study these early West Texas aquaculture pioneers and their efforts, however meager, to grow shrimp and fish in the desert. Then I took real scientific training and modern aquaculture methods and brought the whole thing into a solid, albeit brief, commercial industry and phenomenon whose legacy lasts to this day. This book is not an autobiography or anywhere near the story of my life, but I include a lot about me in this work out of necessity. This was a huge part of my life. I have been called a visionary. Maybe or maybe not, but I certainly drove the whole thing. I started the first commercial farm and owned the last one, and I had something to do with building or managing every shrimp farm that was ever built in West Texas. While I didn't conceive of the idea first, I do believe I did the heavy lifting to make it real. I was the cheerleader, the quarterback, the coach, and the water boy, though many others helped along the way.

I have been moved by the interest and passion of all the people who appreciated and enjoyed this brief moment in aquaculture history. I also felt I owed these people, and those who don't have experience in this story, a real documentation of what happened here and why it was so very unique and interesting—a real tribute to the natural resources of Texas and the impressive accomplishments of its universities and the graduates of those institutions. I hope that you find this story intriguing and that it brings the pioneer and entrepreneur spirit out in all who read it. I hope that in reading this tale, those who were involved or witnessed it look back on the venture with fond memories and those who were not there feel as if they were—or wish they could have been.

SHRIMPING WEST TEXAS

PART I

CHAPTER 1

SHRIMP FARMING 101

I WILL TURN THE DESERT INTO POOLS OF WATER
AND THE DRY LAND INTO FLOWING SPRINGS.
ISAIAH 41:18

I think it useful to give the briefest explanation of aquaculture in general and shrimp farming in particular, for those who are not familiar with the practices—which is most people. Aquaculture is just the method of farming things in water. It is a form of agriculture, and more akin to ranching in most respects. Livestock, including livestock that live in water, is cared for, fed, and nurtured until it reaches a market size and is ready to be sold. The farming of fish has the capability to supply the whole world with high quality fish and plants and to save the world's oceans from overfishing, and it should be encouraged by all. Chances are that most people have eaten shrimp grown in a pond whether they knew it or not. For years now, a very large portion of the shrimp on the American market grew in ponds in Asia or South America. In fact, people have grown fish and shrimp in one way or another for thousands of years.

For the United States and the modern world, shrimp farming all started when Japanese scientist Dr. Motosaku Fujinaga came to the National Marine Fisheries Lab in Galveston, Texas, in the 1960s

5

to describe to the scientists there the life cycle of shrimp, how to make them spawn, and how to rear the baby shrimp. It is all rather technical, but let's just say from there on the American fisheries scientists—especially those from Texas A&M University—took what they learned from Dr. Fujinaga and developed the basis of the industry we know today. Simultaneously, Ralston Purina, the pet food company, took on a major research and development effort in commercial shrimp farming. The company focused most of the development of this commercial industry in developing countries with vast coastlines and where things were cheap and resources plentiful. By the 1980s, South and Central America were crawling with Aggies managing shrimp farms. By 1990, shrimp prices were high enough to get investors interested in more expensive operations like the large shrimp farms forming on the Texas and South Carolina coasts, and the American shrimp farming industry really took off.

The shrimp of choice for almost all shrimp farming worldwide is the Pacific white shrimp *Penaeus vannamei*. This shrimp is native to the Pacific coasts of South and Central America. It is the only shrimp that really tolerates the high stocking densities necessary for a profitable industry and it grows fast on relatively cheap feed. The Pacific white shrimp can also tolerate the wide range of water conditions that might be encountered in a shrimp farm environment. Throughout the years, farmers tried many species of shrimp including the shrimp species native to the Gulf of Mexico, but none have performed in farming conditions as well as the Pacific white shrimp. They are even the preferred shrimp to grow on the vast acres of Asian shrimp farms.

The process, simply put, works like this: the farm gets post-larval (baby) shrimp from a hatchery that spawns them. The farm might supply from its own hatchery or could purchase the larval shrimp from a commercial hatchery. Most American farms do not have their own hatcheries but rather acquire the larvae from other private companies that run dedicated hatcheries whose sole enterprise is producing larval shrimp to sell to grow-out farms. Once purchased and carefully transported to the farm site, the little shrimp are gently acclimated to the ponds that usually have plankton already growing in them as food. As they grow bigger, the baby shrimp are fed dry pelleted food throughout the warm months. The shrimp also get oxygenation via paddlewheel aerators (lots of them) spinning on the ponds, kicking up a froth of water and air and circulating the water to keep it from getting stagnant. After six to eight months of warm weather, the fully-grown shrimp are harvested. Harvest usually occurs in the fall before water temperatures become too cold for shrimp to grow and survive. The farm must then sell the shrimp, usually to processing plants that deadhead and size-grade the shrimp and freeze them for sale into the food service or grocery market. Each step of the process is tricky and contains a lot more detail, but this overview briefly describes most shrimp farming. There is also a shrimp farming method called "super intensive" that involves indoor tanks where all the variables are highly controlled and shrimp can be produced year-round. This type of shrimp farming has not yet proven to be profitable and is very expensive to set up, so it is not very common.

In general, aquaculture—especially shrimp farming—is one of the most interesting kinds of agriculture enterprise that exists, but it is very challenging. Many fortunes have been lost in the aquaculture business, especially in the early years when shrimp farming was still in the experimental stages. What makes this story important and even more interesting is, before the early nineties when we started looking at West Texas as a possible shrimp farming location, no one had ever farmed shrimp inland and far away from the coast using salty well water—at least not in any meaningful way. In most aquaculture and shrimp farming circles, many considered the idea preposterous and perilous at best. But then, I never really cared what most folks thought.

I was trained in aquaculture at the best institutions Texas had to offer and by the best marine biologists and aquaculturists that anyone could study under: Texas A&M and the University of Texas Marine Science Institute, and Dr. Connie Arnold, Dr. David McKee, and Dr. Joan Holt, to name just a few. These folks were legends of aquaculture. I loved my time at the University of Texas Marine Science Institute where I worked and did all my graduate studies with these amazing professors, though the environment felt too safe and too academic for me. I wanted the big leagues, the real world of commercial aquaculture. That master's degree in my hand felt like a sword that I could use to conquer the world—at least the fish farming world. It was also my passport and credential to embark on a grand fish farming adventure.

Immediately after graduate school, circa 1990, I took a job in Florida to run a fish farm. Aquamar was the name of the company, and shrimp and redfish were the species of choice. Aquamar seemed well funded and full of promise, a place where I could turn my academic experience into professional, private-sector experience. I was super exited for the job and to give Florida a try. As it turned out, the Aquamar investors had selected a poor site for this farm, as happens often in aquaculture ventures. They had chosen a location for its aesthetic beauty and access for their large fishing yachts rather than what it had to offer in the way of aquaculture resources. These investors had a vision of fish farming where everything is clean and fascinating. Pictures of clear water and smiling employees adorned their advertising materials. They soon learned that fish farming is more of a dirty job—fun and challenging, but not clean and pretty. Thus, Aquamar investors eventually soured on the business. It was not what they had envisioned or were very interested in once they could see what it was really like. These guys had such a snooty lifestyle that the homeowners' association where they all lived banned traditional wooden Florida skiffs, which are a part of Florida's heritage, because these boats were too ugly and pedestrian for their delicate sensibilities.

These high-rolling investors also got caught up in the real estate and savings and loan crisis of the early nineties, and this took away a lot of their attention and capital. Once the Resolution Trust Corporation came calling, the original group of investors abandoned the project seemingly overnight. I had to scramble just to survive, and after my time in Florida I wanted to get back to Texas where I had people and connections. I felt I could get things back on track and make things happen in Texas.

Luckily, a local doctor from Panama City approached me to help him start an aquaculture venture for his son who had just graduated from college, was interested in aquaculture, and had expressed an interest in growing fish. While Florida was a preferred location, he was not really tied to the location of the venture. How he found me, or this little farm in the woods on the edge of East Bay, I have no idea. I suspect he learned about our predicament on the golf course and at country club parties. Since the location for a shrimp project was no concern to him, I told the doctor of the interesting but slim information I had about a test project in the desert of West Texas. If he would let me go check it out, I could offer my advice and efforts toward his interests.

Just a few weeks before I met the doctor, I learned that a local West Texas water / irrigation district in partnership with the Texas A&M extension service had started work on a demonstration shrimp farm using the salty groundwater from the area. The small site was located down the road from the tiny, dusty town of Imperial in north Pecos County, just up the road from Fort Stockton. They had recently hired a young man named Hector to manage this test operation. Hector was my connection, the one who had told me of the operation in the first place. Hector had been the manager of a redfish hatchery and farm near Collegeport, Texas, and I had purchased redfish fingerlings from him for the farm in Florida during the last couple of years. He heard of the West Texas project and applied for the job of manager. He and his wife moved out to Imperial and lived on the farm. He had only been at the farm a few months when he called and told me I should come and see the potential that was out in West Texas. Hector and I talked often and he knew that I was looking around for a site for the Panama City doctor and his son. He convinced me that I should come and see the amazing area with the cheap land and plentiful salt water. I knew it would be a weird idea for most people to ponder so I prepared for the raised eyebrows and scoffing faces.

I pulled my rental car into the parking lot of a cafe in Odessa, Texas, early in the spring of 1992. This was the place where I was scheduled to meet with Paul, a fellow from West Texas Water Well Service, the premier water well company in the area. I had flown to the area to get as much information as I could, and since this is such a remote part of Texas, I planned on meeting with a whole lot of people and company men. I needed to leave knowing whether to pitch the idea to the doctor/investor I was working for, or just drop the whole thing as crazy. Farming shrimp in the desert, 800 miles from the ocean? I needed a lot of information and data because I would have a lot of explaining to do.

It was pretty early that morning, so I went on into the cafe and ordered a coffee and found a booth. I was dumbfounded when I took a sip. It tasted very strong and bitter and . . . salty! Not just annoying like my grandmother's well water from Spur, Texas, but like the rim of a margarita glass salty. I asked the waitress if the coffee was made with Odessa city water, and she confirmed that it was, but explained that the town would soon get a new water source, and it was supposed to only be about a third as salty. I thought to myself, man, have I come to the right place. They drink salt water here! This location was looking promising for sure. What's with the salt water, one may ask? This is the most important part of how shrimp farming found its way to the area. The West Texas Permian Basin is a unique part of the world covered by ancient oceans on several occasions throughout millennia. The many geological processes that formed the area are described in numerous academic geological textbooks, but succinctly put, due to the high elevation of the Permian Basin, when the oceans receded and sea levels dropped, the basin trapped the seawater in sort of a big, enormous lagoon. This Permian Basin lagoon stretches from Lubbock, Texas, and Roswell, New Mexico, in the north to San Angelo, Texas, to the east and south into Mexico. Throughout thousands of years, sediments would cover up the water, trapping it in subterranean formations. This happened enough times that the vast amounts of trapped seawater formed large salty aquifers, referred to collectively as the Permian Sea.

Many places, usually deserts, throughout the world have briny groundwater, but what makes the Permian Sea so unusual is that this water has the salinity of a healthy bay system—12–18 parts per trillion of salt to freshwater—and marine species thrive in this water in its raw form, straight out of the well. Many other salty groundwaters throughout this country and the world have imbalances in metals or other salts that prohibit marine life without some water treatment or adjustment. The Permian Sea water, however, is good to go just as is. In addition, the salty groundwater in the Trans-Pecos region is only 40 feet to 200 feet deep with flows as high as 4,000 gallons per minute. This means the water is easy, cheaper to drill, and much cheaper to pump. This water resource—this Permian Sea—is the jewel of the desert.

When Paul arrived at the cafe, I told him I wanted to discuss the saltwater situation in north Pecos County. He told me that there was not a whole lot of information on the area as the groundwater there, officially called the Pecos Alluvial Aquifer (the Permian Sea), was so salty that very few wells had been drilled for anything other than rig water for the oil field. Drilling and well information was not available to many. In the 1930s and 1940s some irrigation wells were drilled, but the remaining folks who knew about these were elderly. Paul told me the West Texas Water Well Service had drilled the new saltwater well set to supply the pond water for the little demonstration farm that I was headed to check out.

Paul indicated that for an exploratory well in an area with few documented wells, it was producing steadily and providing the quantity and quality of water the test farm needed at the time. He also added that he believed they could do much better in the future by drilling simpler wells like the farmers drilled for crop irrigation rather than the over-engineered well they were required to drill for this project.

As I would later find out, the little test farm was basically a project started by two government entities, Texas A&M University and the Pecos County Water Improvement District (PCWID). While it was a noble and useful endeavor, many unnecessary expenses and formality went into the venture, as is characteristic of government projects. Four PhDs, three engineers, and a scientific advisory board planned every little step in the project, which slowed down the development and introduced inefficiencies along the way. I remained encouraged by the description of the water resources in the area, and I formed a connection to a water well drilling company that could get me the most important thing of all for this project—beautiful, pristine, clean, virgin salt water with ten to fourteen parts per thousand salinities. This was like perfect bay water where larval shrimp live and grow to adults in nature. I was getting excited about the whole idea.

Still, I had to temper my ever-increasing desire to return to Texas with a sober look at the facts. I knew if I ignored any red flags and did not heed any warnings, I would find myself once again in a situation where the site selection was all wrong and the venture would not have a prayer to survive. In addition, my wife Patsy loved my choice of career, and she loved the ocean. North Florida was right up her alley. She was a mermaid and a beach bunny who lived at the beach. She swam, fished, and collected scallops by the sackful. She even captained a little crab skiff for a soft-shell crab operation for a while. To say moving to the desert of West Texas would be a challenge for her would be a major understatement. I knew it had to be right, or this could also be a disaster on a personal level. I finished breakfast, polished off the last salty cup of Odessa coffee, shook the hand of my new friend Paul of West Texas Water Well Service, and headed fifty-four miles south to the grand metropolis of Imperial, Texas, population 420.

Traveling along Interstate 20 through the Permian Basin really brought home the vastness of the oil patch in the area. Oil well after oil well as far as you can see. Then when you turn off the highway onto Farm Road 1053 you're smack in the middle of sand dunes that will rival anything the coast has to offer. The dunes go for miles and miles, relics of the time when the area was an ancient ocean, and these impressive dunes were the beach. Eventually the dunes flatten into desert scrub land with distant mesas on the horizon to the south. When I crossed the Pecos River into Pecos County I arrived at my destination, with a lot of questions in tow.

Imperial turned out to be a most unimpressive desert town. The school occupied the largest area of town, and the library and the public swimming pool complex were a close second. As I drove through the whole thing in the blink of an eye, turned around, and then drove through again I saw one gas station, one cafe, and very few houses—most of which looked like they were falling down or abandoned. Surely, I thought, there is more than this; the main drag has got to be somewhere around here. Nope, what you see is

what you get. I think the sign for population 420 was very gener-
ous—maybe someone was counting the good folks resting in the
cemetery. My contact Hector told me to find PCWID No. 3. Of
course, no address was provided since no one from the area used
street addresses to locate anything. The town was so small, people
just knew where everything was and assumed any visitor could easily
find whatever they were looking for, which was normally the case.
Although I had read that years ago there was a lot of farming in
this area, when I crossed the Pecos River into Pecos County, I saw
no sign of any farming and little sign there ever had been any. This
made me assume the folks at the water improvement and irrigation district would have plenty of time to get me caught up on the
situation—they would not be busy irrigating any farmland, that's
for sure. After my car tour around town, I accidentally walked into
Water District No. 2. Whatever became of Water District No. 1 is
still a mystery, along with the question of why land with no farming had more than one irrigation district. The secretary to Water
District No. 2 seemed bored out of her mind and wanted me to
sit and talk, probably just to stave off the tedium. I apologized for
disturbing her sleep and eventually found my way across the road
to PCWID No. 3. Finally, I was in the right place, these were the
folks who partnered with the extension service and the university.
These were the folks growing shrimp in the desert—or were about
to give it a try, anyway.

At the office, I met a very nice secretary by the name of Gayle,
the only full-time employee of PCWID No. 3. She was a local gal
who had grown up in Imperial and never left. While it turns out
there actually was a very small bit of farming here and there, it
was not enough to support full-time irrigation district employees,
except for the clerical and office work. It is a government entity
after all, so you know there are stacks of paperwork to fill out and
forms to file. They did have a couple of part-time employees, and

I learned that PCWID No. 3 also took care of the other irrigation district's (PCWID No. 2) maintenance and operations when they got the opportunity to irrigate on a few occasions. These water districts usually hired a ditch rider on a part-time basis, and some high school kids when they needed help when irrigation water was sent to the fields. The ditch rider job was not a highly sought-after position—they told me the job had a bad history of altercations with farmers. The ditch rider was responsible for getting the water to the farmer's field and regulating the amount of water based on what the farmer purchased from the district. The ditch riders' calculations were often suspect, resulting in various levels of tension in the field. Sometimes the farmers didn't purchase enough water to complete a watering due to miscalculations on their side, and disputes often occurred. On occasion, the arguments regarding the water would turn physical. Most agreed the ditch rider was not paid enough to take an ass whipping, so the job often remained vacant, and the kids or the farmers ran the water on the honor system.

These irrigation districts and their formation have a history dating back to the turn of the twentieth century. Just twenty-five years prior, the Comanche were still scalping anyone foolish enough to try to settle in the area. By 1900, speculators from back East started irrigation companies and advertised available land to would-be farmers throughout the Midwest and eastern United States. The irrigation companies took out ads in newspapers in the North and put out fliers in parts of the South where Reconstruction had folks looking to move West. The problem was that the water was salty, so these land agents and promoters would go to great lengths to keep this no small issue from interfering in their land sales. The land speculators would obtain fresh, drinkable water from outside of the area and bring it to Pecos County. They hid containers of the fresh water in strategic locations along the Pecos River, where they would stop with a potential client and, out of the client's sight, go fetch some "river water" and put on a grand display of how sweet and fresh the water was and explain how the land was just ripe with potential. They had the sales pitch for sure. Cheap land, no more Comanche, railroads through the county, and new roads and towns being built every year. In this new century, northern Pecos County was going to be a paradise . . . and in 1913, six Imperial area farmers committed suicide.

According to old timers and the historical and immense set of books titled *Pecos County History* sponsored by the Pecos County Historical Commission and edited by Marsha Lea Daggett (Staked Plains Press, Canyon, Texas, 1984), the crop failures were too much to bear. People gave up everything back East to come West. They spent everything they had and came with the hopes and dreams and desperation of a new beginning. The truth of this area was devastating for all but the hardiest and most stubborn settlers. These irrigation companies built these vast irrigation systems, but good, fresh water seldom flowed in the Pecos River, the source of the water for irrigation. Most of the growing season the river was salty, and putting salt water on the soil only lowers its ability to grow anything. It is important to understand how difficult it is to farm on salty, alkali ground in a place where the average rainfall is twelve inches per year and where temperatures often get to 115° Fahrenheit in the summer and -10° Fahrenheit in the winter. Almost no crops of any value can grow in this mess, and the ones that do have very poor production.

If the salty dirt doesn't suck the life out of the plants, the super-hot, dry desert air will. It sucks the life out of the people too. Most settlers either left for farther West or went back East where they came from. Some farmers, reduced to poverty by the harsh conditions and with no way to leave, had no choice but to continue. As time passed, some of them learned how to farm a few salt-tolerant crops like grass for hay, cotton, and sorghum. It was a pretty tough existence. By the end of World War II, most farms had been abandoned or repossessed by the banks that could not give them away. A good bit of the land became grazing land for cattle, but most folks went to work in the burgeoning oil industry. Some of the very first oil and gas wells in West Texas were drilled in the Imperial area. The water districts, which started out as private ventures at the turn of the century, eventually became public, taxing entities run by local, elected boards. The allotments of water in the Pecos River watershed were, after many years of litigation between districts and between the states of Texas and New Mexico, adjudicated with certain water allotments going to each district. More importantly, though, a nice sum of money was paid by the state of New Mexico in a settlement to the state of Texas in an agreement called the Pecos River Compact. The money is allocated to each district based on the irrigable land in that district. The context for the existence of the water districts sets the stage for how aquaculture started in the area. The water districts had very little activity due to the lack of farming, and justifying the need for an irrigation district was a hard argument to win. The area had few farms to water but a big pile of money in the bank dedicated to improvements for water projects in the area—money that could not be used for any other purpose as mandated by law. The water districts could only mow and grade and maintain unused irrigation ditches for so long before the state of Texas started asking if the districts were necessary at all. After all, those Austin politicians would love to get their hands on that money.

The board of directors for Imperial's PCWID No. 3 had forward-thinking board members in the early 1990s, and probably the most imaginative person on the board was Ernest Woodward. He was a native Pecos County cowboy from a well-established local ranching family. Back then, Ernest was the president of the PCWID No. 3 board of directors. He is a born leader and true man of the

West. He was one of my favorite people the whole of my time in Pecos County. He knew that traditional dirt farming would never be any real industry in the area due to the condition (salinity) of the water, but he did have a feeling that aquaculture might just be able to thrive. Now this was not something that would directly benefit the water district since shrimp farms would not be using irrigation water provided by the district. They would need the saltier well water from the underground aquifers in the area, but he knew that if an aquaculture industry could be developed it could be an economic boost to the area in general. Ernest's intuition didn't come only from his gut, it also came from his knowledge of the results of a few attempts by area men who tried to experiment with fish and shrimp farming.

The very first attempt at aquaculture in the area happened in 1958 when the Texas Parks and Wildlife Department stocked redfish fingerlings in Red Bluff Lake near Orla and in the Imperial Reservoir. Then in the late 1970s the local ag extension service was able to obtain some redfish (*Sciaenops ocellatus*) fingerlings and post-larval shrimp from the state and put them in a couple of Imperial-area gravel pits that were flooded with brackish groundwater. They ended up harvesting about a hundred pounds of shrimp, and the fish lived and grew to several pounds, though they eventually died from the very cold West Texas winter temperatures. Still, it was promising that marine species could live and grow in the briny, local waters.

Several years later in the late 1980s there were two, much more substantial attempts at growing shrimp in ponds in West Texas. One of these was in Martin County up near Big Spring and the other was in Crockett County in the vicinity of Iraan, Texas. In both cases they were ranchers who scratched out a couple of one-acre ponds, pumped the brackish well water into them, and then stocked with larval shrimp that they purchased from a hatchery in Harlingen,

Texas. They might have had a little direction from the extension service, but it was probably limited as most farmers and ranchers I have met in West Texas give very little regard to the advice of the extension service in general. I suspect they flew by the seat of their pants on these projects. The thing that was promising was that they got more than a thousand pounds per acre of medium-size shrimp that sold at a nice price in their local areas—proof that even a blind pig can find an acorn once in a while. I am sure these ranchers were experts with four-legged animals, but I am fairly certain they knew absolutely nothing about ten-legged animals and thus were in way over their heads. There are some things all livestock operations have in common, but in general, aquaculture is completely different from cattle ranching. Neither of these operations lasted more than a year or two and were more of a novelty than anything else because those ranchers were about as successful in raising shrimp as I would have been raising cattle. I think the biggest reason we would eventually become successful at shrimp farming in Pecos County is because I have a graduate degree in marine biology with a specialization in intensive shrimp farming. I certainly obtained a lot of on-the-job training and experience that would become the backbone of my career, but my formal training was the glue that made it all stick together.

However brief, the efforts of those two operations caught the attention of people in the whole of West Texas, not the least of whom was Ernest Woodward. Ernest had a feeling that this just might be the sort of thing that, if done correctly, could be a game changer for the area along the Pecos River. Ernest and the board of PCWID No. 3 sought out the help of Texas A&M University, which was a good move and ensured that, in the beginning at least, the project would have technical support from the very university that had conceived the idea of shrimp farming and was responsible for its global development. The district then commissioned a report called "The Feasibility of Aquaculture in Pecos County and Far West Texas" completed by Durwood Dugger in May 1991. It was a report that contained the experiences of the above-mentioned pilot projects, information on the resources of Pecos County, and some economic feasibility modeling data around building and running a commercial shrimp farm. This last bit was far removed from anything realistic and of no use whatsoever. It was based on building and operating a theoretical 200-acre shrimp farm in West Texas. When using well water at 3,000 gallons per minute, a one-acre pond three feet deep is a lot of water. The feasibility of running 200 acres of ponds in this part of the world was very low, unless the operation was bankrolled by Jeff Bezos or the United States Department of Energy! The largest shrimp farm ever built the entire time aquaculture was in operation in West Texas on was my farm, the Permian Sea Shrimp Company, and it was seventy-five acres of saltwater ponds.

Logistical shortcomings aside, the little report was useful and a good starting place for a would-be shrimp tycoon, and I got three copies of it from the secretary at the district on my first visit. She was excited that someone finally wanted a copy.

THE DEMONSTRATION FARM

What the study and feasibility report did accomplish was to convince PCWID No. 3 to build a demonstration farm. This farm would serve two purposes: to grow shrimp in a manner consistent with the current science and commercial practices, and to attract potential investors and farmers and give them real data and information that would be useful in helping them start farms in the area. They were to do this with the help and advice of Texas A&M Extension Service and some of the research personnel and students from the Wildlife and Fisheries Science Department of Texas A&M. The plan was to build ten one-acre ponds, two 2.5-acre ponds, and a few little nursery ponds. Some greenhouse covered nursery raceways would be built also. The greenhouses consisted of small, rectangular depressions in the ground, lined with plastic liner material with greenhouse plastic held over them by fiberglass oil field sucker rods. The purpose of these nursery raceways was to get the baby shrimp a little head start on growth in the early spring when the weather was still fairly cold in West Texas. This way they would effectively lengthen the growing season by starting shrimp safely indoors when it was still too cold for them to grow outside. The young shrimp would then be moved outdoors to bigger ponds once the weather was warm. Either two crops a year of shrimp or larger shrimp in a single crop was the expected outcome of this head starting procedure. All of this was under construction when I made my first visit to Imperial.

Before I headed out to see the future shrimp farm site, Gayle put me in touch with a local contractor by the name of Ronnie. He had some earth-moving equipment and ran a little company that did all sorts of dirt work and contracting. He had helped build the ponds for the test farm, so she thought Ronnie would be a good contact for me. As it turned out he was also on the PCWID No. 3 board of directors, and I ran into him in the parking lot just as I left the office. He was a super nice guy and indicated that his company did mostly oil field work, but that he could also build pits, as they are called in the oil field drilling business. He told me that he was also a rancher and didn't really understand the whole shrimp farming thing. He did hope that it would work and be good for the area. He also had some sage advice for me when he said, "I hope you make a lot of money with this, but if you do, everyone in Imperial will hate you. It's just the way they are. Then again, if you don't make money, you'll never be able to afford to leave this place, and someday, that's something you're definitely going to want to do!" Those were prophetic words of wisdom.

Next, I left the water district office and headed to the demonstration farm to meet up with Hector and get the lowdown on the potential for shrimp farming in the desert. I drove about three miles west of town on Farm Road 11 and into the gate of the project. Hector had a lot of enthusiasm about the possibilities that existed, but the program had yet done very little along the lines of growing much shrimp. The project was very new, and it was also very early spring when the temperatures were much too cold to grow shrimp.

They had apparently played around with shrimp and redfish in some of the little nursery ponds the summer before, but they were gearing up for a real go at production in the coming spring and summer. I was a bit disappointed in the fact that I could not go back to my investor with any promising information. It was just too premature at that point. I was damn sure not going to recommend to my investor that he spend his money on a venture in the area with the paltry information Hector could provide at the time, nor could I take the risk of relocating to Imperial only to find out the operation was just a pipe dream.

The demonstration project did have some shrimp in the nursery raceways under greenhouse cover, and they were thriving. At least I could trust that the shrimp would live in the water here, but that information was expected. I toured around the farm, had a good inspection of the empty ponds, and met Hector's project crew. Rosie Montoya was a lovely older man and probably one of the nicest people I ever met with the exception of maybe his wife and his son Danny, also a crew member, who would become a good friend of mine for many years. Last was Manuel Jurado, whom we called Big Manuel. Big Manuel was very skillful at running heavy equipment, especially a backhoe, which is probably the most useful piece of large equipment on a shrimp farm. Manuel was a friendly and reliable hand around the farm. He lived in Imperial with his wife and son, also called Manuel. We called the son Little Manuel, though he was quite a bit larger than Big Manuel. Although Little Manuel (or Manuelito, as he was sometimes called) did not work for the water district, he would later become my right-hand man for all the years I was in the shrimp business in West Texas. This farm crew was plenty capable to work the little operation and did a fine job with the little technical direction they got.

Unfortunately, Hector was more of a fish man than a shrimp biologist, and the assisting academics from the university were not skilled at conveying information to the practical crew of farm hands. I could tell this whole project needed some help. I was also in need of a whole lot more information and data than I was able to obtain on this trip. The project was just at its infancy and did not have any serious production attempts from which to obtain good, reliable information. I had nothing to take to my investor to convince him this was the place to be. I wanted this to be the place, but I had nothing but hope and a gut feeling—neither of which are reliable business strategies. I met with Hector to discuss my dilemma, and after much talk we decided that it would be a good idea if I came back in the late spring to help with the first production runs in the big ponds. He felt he could use my help and experience in putting in the first crop, and I would be able to get an eyewitness account of the whole situation from start to finish. I could also start looking for a site in the area and be ahead of the game if the production was promising.

I decided to call the doctor and asked for more time. My plan was to come back and live in Imperial for that next summer, help with the demonstration farms' first shrimp crop, and get the data I needed at harvest time. I would need till August or even September to gather enough information to make a recommendation. In the meantime, he and his son could still look for property in Florida to build a shrimp farm, just in case the West Texas operation turned out to be a flop. I was not terribly excited about either option at that point. I did not have enough information to get excited about West Texas, and I certainly was not excited about looking for land in Florida, which is sold by the square foot on the coast. It is phenomenally expensive, and the regulations in Florida for aquaculture were unbearable at the time. No one on the Florida coast wanted a shrimp farm in their neighborhood—any place seafood is produced for commercial purposes is seldom pleasant on the eyes or nose. Fish processing plants were common in most Florida coastal towns but were relegated to the harbors and waterfronts where the unsightly sights and smells (and the inevitable horde of pooping seagulls) are considered quaint to the tourists who like to go to see that sort of thing for about 15 minutes then go back to their rented, picturesque beach houses and order pizza.

After returning home, briefly, to Florida, I did a bit of site search-ing in the local area and soon decided that Florida was not going to work for a shrimp farm. My wife and I also decided that we didn't want to stay in Florida any longer and made plans to move to Texas. Maybe it was the dry desert air or the time spent back in Texas that pushed me over the line. Patsy and I packed our belongings, rented a U-Haul, and headed back to Texas. I informed the Panama City doctor that I would bring him the results of the summer produc-tion and make my recommendation in or around September. After making the long drive west from the Florida Panhandle, I dropped my wife and new baby off at my parents' house in North Texas and headed back to Imperial. I did not have the nerve to take my wife out to that desolate place just yet. I knew I needed strong justifica-tions for moving our family to that dry and dusty place. It would not make any difference what the investor wanted to do if I didn't have a good, solid sales pitch for Patsy. I needed that production data in a bad way.

I was happy to be growing shrimp again and even happier to be doing so in a unique place. So much potential but with a large and interesting challenge: making a bay-type ecosystem that would support marine life way out in the desert. For a young and confident marine biologist, it was the chance to really do something novel and original—go where no man had gone before and all that. No one was more up to the task than me. I had been around agriculture my whole life and always wanted to have my own farm. After see-ing all the expenses and roadblocks for doing aquaculture on the coastline in Florida and Texas, I knew that without a vast fortune the location would have to be where the land was cheap, the water plentiful, and where no one cared about the view. I was the kind of guy who always wanted to blaze a new trail, to do what others said was impossible and to make my own rules. It was possible that this would shape up to be just the place for me, and possibly a new and lucrative shrimp farming industry.

There were no motels or places to lodge in Imperial, so I stayed on the demonstration farm in the twenty-foot travel trailer provided by the water district. I suppose they had the RV put there specifically for accommodating folks who came to see the project. It turns out that I was the only one that ever used it. Hector lived in a mobile home on the farm as well. His lovely wife Sandra was a great cook and took good care of me. It was beneficial to sort of camp out in this new environment and get a feel for the whole place on a primitive level, living and breathing in the desert with few distractions. Dry air, warm days and cool nights, stars and more stars like you have never seen anywhere else. Lots of migratory ducks flying through and the coyotes howling all night, every night. Badgers with their enormous holes in the ground, porcupines and rattlesnakes—lots of rattlesnakes. And we can't forget the scorpions. These you might find anywhere, including your bed! There was also the smell of West Texas. The aroma comes from two major sources: the hydrogen sulfide from the oil fields and the strange smell of alkali soil, especially when it is damp or wet after a rain. The combination creates a unique West Texas fragrance. The dry air amplified the sharp smells in your sinuses.

After I was back in Imperial and settled in, it was not long before the weather started warming up and it was time to get the full operation going. We stocked the larval shrimp that had been head started in the greenhouses out into the one-acre ponds for that first summer. The stocking level was somewhere around twenty-five per square meter or so. The little guys were about a half a gram to a gram each at stocking, which was a very nice size with which to start production ponds. We stocked six one-acre ponds. We put in each pond two paddlewheel aerators of three horsepower each to churn up the water and keep it oxygenated. This oxygenation step would be imperative in the late summer when the weather would be very hot and the shrimp biomass in the ponds would, hopefully, be high. I knew it would be hot, but I really didn't have a clue just how tortuous it could be in this part of the world. Warm water holds very little oxygen, and shrimp need a lot of oxygen to thrive. A large amount of shrimp needs an immense amount of aeration. We needed twice the aeration that we started with, but at least we had the extra aeration equipment on the farm as the district had purchased quite a number of these paddlewheels. This turned out to be fortuitous.

The project had also purchased a little, galvanized blower feeder from Idaho, originally designed for feeding trout. Typically, when feeding fish or shrimp in ponds one uses a device to blow the feed pellets out into the pond. The device is usually mounted on a trailer or on a truck so that you can drive around the pond blowing the feed out into the water. The shrimp wouldn't care how the feed got into the pond as long as it got there at least four times a day. Shrimp eat all the time, so for good production, feed must be presented often.

To grow shrimp successfully takes a lot of work. Not hard work, but there are tasks to be completed constantly. Feeding several times a day is just part of the work. Oxygen levels in the water need to be checked often and general water quality tested each day. Farmers collect samples to determine the health of the shrimp and to weigh and measure the shrimp to keep up with growth rates. It was a good thing that I was at the demonstration farm that important and critical summer to help them with all these tasks. I had not been out of graduate school all that long, but I had been managing the

Florida fish and shrimp farm the last three years and was around commercial aquaculture farms since college. I never tried to act as the manager, but it seemed to give the Imperial crew some confidence that we were doing things right since someone in leadership actually had commercial experience.

That summer holds good memories for me; I really enjoyed meeting some of the folks in Imperial and spending time with Hector and his wife. I remember once she was complaining about how ugly and stark the desert was. It was spring, and I had noticed a number of beautiful flowers growing here and there in the vast landscape around the farm. I picked her a nice bouquet of pretty flowers of all colors and kinds. When the glorious prize was presented to her, she gasped and asked where they came from. I told her they came from all around us here in the desert—you have to look for it, but the beauty of the desert is right in front of you. I don't think she had the same attitude about the area after that.

I also spent a lot of time with Big Manuel and his family that summer. He invited me, the sole gringo, to a pig-butchering one weekend. What an experience that was! Through the years I went to many of these butchering parties, but this was my first time and I found it very enlightening and a real hoot. It went something like this: the pig was selected from the pens and shot in the ear with a .22 cal. Then it got a more thorough bath with soap and water than I had gotten all summer in my little RV. Boiling hot water was poured over the carcass, which caused the hair to slip. The hair was removed until only a very smooth, lily-white, hairless pig remained. Then the thing got bath number two with soap and water again. This was necessary as the skin was going to be the party food, along with copious amounts of beer and spirits. You simply can't have anyone hungry or thirsty while the butchering was going on. That is very bad manners.

They pulled out a "disco," added vegetable oil to it, and put it over the fire. A disco is a plow disc with the center hole welded shut and some handles welded on—like a wok but much bigger and heavier. It is the signature cooking device of the desert southwest. Once the oil was really hot, they cut off strips of the pig skin with some underlying meat and pitched them into the boiling oil. These *chicharrones* fried up on the spot with the little strips of meat still attached are about as good a party snack as you ever ate. The merrymakers continued to cook and eat but now they turned their attention to the next important step: gutting and butchering the beast, just as anyone would butcher a pig anywhere else in the universe. No magic to it, just a great party. The meat butchering at this particular party, as in many others I was able to participate in, extended into the next day owing to the vast amounts of alcohol consumed which caused the group of reveling butchers to lose focus. They would fall out, one by one, here and there.

When it came to the drinking, I certainly was right there with them, but rather than pass out on site I usually managed to talk my hosts out of a nice hunk of pork and a sack of the *chicharrones* before I stumbled my way back to my truck and snuck back to my little trailer on the farm. Provisions for a week! These are the sorts of things that make living in the rural Trans-Pecos desert something interesting and fun. In an area where there are not a lot of people, socializing and entertainment comes in the form of barbeques, church potlucks, hog butchering, and the like. Hell, some folks do day work for ranchers gathering cattle, just for the camaraderie with other folks . . . and the food, of course. Ranchers feed day hands pretty good. Such was the life in Imperial, and I started feeling truly alive there.

By August, we were ready to harvest the ponds. The shrimp were medium to large tails and weighed more than twenty grams each. The cast net samples were showing good numbers in every throw. The fun was over and it was time to work. Shrimp harvesting is work—period. It can be fun, but it is serious hard work, mostly done in the cool of the night as an insurance measure against heat-related spoilage. It takes hours and hours to complete a proper pond harvest, especially in a remote and challenging location. Thousands of pounds of ice must be obtained the day before harvest, totes set up to ice down and preserve the shrimp as soon as they are out of the pond, equipment readied, customers or a processing plant arranged, and transportation to market secured. Mistakes are very costly in an area where the temperature might be 110° Fahrenheit, so everything must be organized well, or you end up with thousands of pounds of rotten, smelly, unsellable shrimp.

We were also expected to inform the game warden that we were going to harvest. Originally, shrimp farming was a coastal enterprise along the Gulf of Mexico. The law requires farmers to inform the local Texas Parks and Wildlife Game Warden when draining ponds full of exotic shrimp so they could inspect, at their discretion, and determine whether or not the shrimp could possibly escape into the natural waters of the state. Any escaped exotics could be, and often are, a big problem for the native population. Think Asian carp in the Mississippi River or Burmese pythons in the Everglades, or New York Yankees fans in Houston. This "call the game warden" rule was a bit silly for a shrimp farm in West Texas since no shrimp were going to escape to the Gulf of Mexico eight hundred miles away. But it was a bona fide government regulation. It also didn't make any sense to the local game wardens, who told us they were not going to come to Imperial at midnight and inspect a shrimp harvest. These dry desert game wardens probably thought we were just pranking them anyway. The law didn't say they had to come, it just said we had to call them.

THE VERY FIRST HARVEST

A shrimp harvest is just like a rock concert—hot, sweaty, and loud. If the band is well prepared, it is a joyous occasion with hearts a glee, but if the band is having a bad night, then it is just a bunch of noise with mad, smelly people who all wish they were somewhere else. It takes all night from sundown to daybreak plus a good portion of the next day to finish a pond harvest, and that is without any problems or issues to slow things down. To start a pond harvest, water is slowly drained from the pond in a manner that allows for the shrimp to follow the water out of the pond and into the nets or pumps without stranding on the muddy bottom. If the shrimp get stranded on the mud, or as we called it "stick," then they must be picked up by hand, which is a job that *really* sucks. There are always some shrimp that stay in the pond and must be retrieved by a team of pickers, but the idea is to keep this number to a minimum. The picked shrimp must be washed of the black pond bottom mud and all the shrimp put on ice and cooled down as fast as possible. The whole procedure is not pleasant work, but it is fun in a weird way.

We learned a lot from our first harvests on the demonstration project. The coastal farms had started using a harvesting device called a shrimp pump that catches shrimp as they come out of the drain then pumps them up to a screen system where the shrimp can be collected in shrimp baskets and dumped into the ice. It was quick and easy, but the shrimp pumps were expensive. We didn't have the money for such fancy and modern equipment, so we had to go old school with the live car system—a 15 feet x 8 feet x 4 feet deep net box with an open top that acts as a trap for all the shrimp coming out of the drainpipe. The pond drainpipe goes through the levee and into the drain basin outside the pond. The live car net is tied to the end of the pond drainpipe which is submerged in water inside the drain basin. When the pond is drained, and the shrimp follow the water through the pipe into the live car net box, they remain covered in water—like a big net cage for the shrimp. This keeps them alive and fresh. The shrimp then must be manually dipped out of the live car with large nets and dumped into a boom net affixed to a crane or backhoe boom. The backhoe operator will then move the boom net over to the totes, filled with ice, on the levee and dump the shrimp. The totes hold roughly 1,100 pounds of shrimp and 400 pounds of ice and water. This is a lot of work. A hand net might only hold ten pounds of shrimp and four or five people have to work fast, constantly dipping shrimp out of the live car into the boom net. Quality is the main goal, and you don't want shrimp getting hot. The shrimp need to hit the ice alive if you want the best product. In the years to come we would improve harvest to the point of precision, but at this point in the development we just had to do the best we could using very crude and simple technology like the live car net.

Local volunteers often helped us with harvest. For this first harvest, there were quite a few folks who wanted to help as this was a very new experience for most in the area. I noticed, however, that many of the volunteers and even some of the paid crew were useless around the water. Harvesting shrimp requires getting wet and muddy, but some of the locals acted as if the water was sulfuric acid or something harmful. They just refused to get it on them or get near it. I guess living where there is almost no water makes folks averse to the stuff. I thought it strange, but during harvesting we didn't have time to think much about it. We just had to push through with the hands that were willing.

We managed to harvest all six ponds in the first part of that August by doing about three ponds a week. Some ponds just took a night and a morning to finish, and a few troublesome ponds took longer to complete due to shrimp stranding. These experimental ponds had very flat bottoms that did not drain out well, so if we had a trouble pond and could not finish by early daylight, we would get the shrimp we had collected in the live car into the totes and then re-flood the pond with about a foot of water. The next night we re-drained the pond and gathered the rest. In the end, the shrimp were beautiful and healthy looking, but best of all the production was 3,000 to 4,500 pounds per acre. Even though later in the 2000s we harvested close to 10,000 pounds to the acre regularly, at the time these numbers were on par with what farms on the coast and in South America were getting. I knew without a doubt that this could be commercially viable. In addition to great production numbers, we got some good economic numbers as well. We put some local labor together, de-headed a bunch of the shrimp, and sold them all across Pecos County and in Odessa as well.

On one of the first trips selling shrimp in Midland we set up in a parking lot just down the block from a restaurant. We were very busy that day as many people came to purchase the shrimp. Eventually we caught the attention of the family that owned the restaurant. They bought some shrimp from us and took them back to their eatery. In about thirty minutes, they returned wanting more. I was so proud and bragged to the folks from the water district about how the shrimp were going to be in a restaurant in Midland, Texas, and this was just the beginning of big things. Well, the family that owned the restaurant heard me and said, "Oh, no sir, this shrimp is way too good for us to buy to serve in our cafe. We are buying these to take home with us to eat at our house!" I guess that was a compliment. Anyway, people loved them and paid top dollar for them everywhere we set up and sold them. I don't know if it was the water or the climate, but that shrimp was without a doubt the best I ever ate. Everyone else seemed to feel the same way. Now I had the data I needed to make the pitch to the doctor/investor: good production numbers and proof the shrimp would sell. I also needed to explain to my wife that the path to our hopefully very bright future just took a sharp turn to the Southwest.

North Florida waterfront dock with Patsy's crab skiff.

Ponds at the PCWID No. 3 demonstration farm.

Harvesting at PCWID No. 3 demonstration farm using the live car system.

Moving shrimp from the live car into totes with ice.

First shrimp de-heading operation at the Water District (PCWID No. 3).

Mennonites building ponds for the Triton Aquaculture farm.

Triton Aquaculture farm.

Picking up shrimp stuck in the pond at harvest.

Lost Sea®

West Texas Desert White Shrimp

Five hundred miles inland from the Gulf of Mexico, deep beneath the northern Chihuahuan desert, flows a hidden sea. Covered for eons by the desert sands, the Permian Sea contains some of the purest saltwater on earth. This pristine water provides the growing environment for Lost Sea® West Texas Desert White Shrimp. The result is a premium sweet white shrimp like nothing you've tasted before.

THAWING

For best results, thaw shrimp overnight in the refrigerator. Puncture bag and place in a collander. Put a shallow pan under the collander to catch any excess moisture. For quicker thawing, place unopened bag under cold running water.

Capture the sweet, natural taste and succulent texture of Lost Sea® West Texas Desert White Shrimp with this simple recipe:

Thaw shrimp in refrigerator overnight. Peel shrimp (leaving the tail and last shell segment intact); rinse with cold running water and pat dry with a paper towel. Thread shrimp on skewers and brush lightly with olive oil or butter. Place skewers on a broiler pan 4 inches from heat. Broil until shrimp are pink and cooked through, turning once—about 3 to 5 minutes. Serve with lemon and enjoy!

Ingredients: Shrimp.

Grown in Imperial, Texas.
Distributed by: Triton Aquaculture, Inc.
2414 Pretty Bayou Drive • Panama City, FL 32405

Bags used to package Lost Sea shrimp for market.

GOVERNOR
AnnRichards

August 26, 1994

Mr. Bart Reid

P.O. Box 448
Imperial, Texas 79743

Dear Mr. *Bart*

Thank you so much for the "home-grown" shrimp. This will be my first opportunity to sample farm-raised shrimp from West Texas and I am anxious to try them.

Please continue with your efforts to bring innovative and progressive economic development to your part of our great state.

Again, many thanks for sending a sample of the bounty of Texas.

Sincerely yours,

Ann

ANN W. RICHARDS
Governor

What a great thing. I always knew we'd find a good use for all that saline water under West Texas!

a.

Letter from Texas Governor Ann Richards.

Aerial view of Regal Farms shrimp farm.

Feeding Regal Farms shrimp with a hopper trailer.

PART II

TRITON AQUACULTURE COMPANY

The doctor and his son went for the plan. I won't say I am a great salesman, but it is likely they could see the optimism in my eyes. The potential was tremendous, and the son liked the idea that we would be doing something new and exciting that no one had ever done large scale, away from the coast. The doctor was a little hesitant until I explained that thirty acres of shrimp ponds would be an order of magnitude cheaper in West Texas than anywhere else, especially coastal Florida. For the frugal doctor, that was the clincher.

I already had a piece of property lined up for this venture: a 120-acre parcel located on the Pecos River adjacent to the water district project. I thought it would be a good idea to stay in the immediate vicinity of the project where I learned from the previous summer's experience that the water and clay soils were suitable for shrimp farming. I would learn through time that there are several other really good locations in the Imperial area with even higher water flow rates and better clay for pond soils, but at this point I was still a greenhorn and didn't want to take any chances or make any big mistakes. The site of a farm is almost always the single most important piece of the deal. I have seen really good aquaculture companies fail because the site was not chosen well. A great site can often help overcome bad operational mistakes. The most important aspect of the site is that it contains all the suitable resources necessary to successfully operate. Choosing a site because it is cheap, you already own the land, or it has a nice view are all bad reasons. No matter how good the view is, it's going to smell like a fish farm soon enough anyway. I could check all the necessary boxes on the site I found, and I endured some weird experiences to obtain it.

During that summer while I lived in the trailer at the demonstration project, I started looking around for a possible site for a commercial farm just in case everything worked out like I hoped. I am the eternal optimist for sure. I figured that if I found a good site, I wanted to be ready to move fast. Some locals had told me that a local independent oil man—whom I'll call Carl—had some land available near Imperial and was willing to make some sort of deal on it. I asked around and snooped a bit through county records until I found where the land was located. Once I learned where it was, there would need to be a little reconnaissance trespassing on it to have a look, just to see if it was at all suitable. It did have high potential. As I walked all over the property, I noticed it wasn't just close to the water district farm but directly adjacent to the demonstration project. We would be neighbors, which was good—less risk since we know there is good, salty water and clay soils right next door. I went back to the water district and asked where I could find Carl. They told me that he had breakfast every morning at the local cafe. Since there was only one cafe in town, I figured it would be easy to find him. What I didn't know was that Carl was sort of a local celebrity, and celebrities tend to be difficult to meet.

I went into the cafe one early morning in May 1992 and hailed him. He did look like a celebrity with his own groupies. Various hangers-on, including a number of oil field pumpers, surrounded him. A short, fat, mean-looking man motioned for me to sit at an empty table and mumbled in a deep, gravelly voice to wait until he was done. Carl must have gotten word I would be coming. It was at least an hour before he was done eating, telling wild stories and chain-smoking at least two packs of cigarettes. Once his entourage had left, he finally acknowledged me. I told Carl that I understood he owned some land he might consider selling, and that if he had time, perhaps he could show it to me and discuss a price. He said he knew who I was and that he had heard I would be looking for him but explained that he didn't have time to show me the land that day. He asked if I would come back to the cafe the next morning, when he might have time. This exact same scenario went on for four days. I would come to the restaurant, he would ask me to wait, and after an hour or so, he would tell me he did not have time to show me the property. At least I got smart enough to start ordering breakfast during the wait. By the fourth day I told him, in a very agitated tone, that I was going to the Fort Worth area to visit my wife for the weekend, that I would be back in Imperial the next Monday, and I would be in that cafe on Tuesday. I wanted to see that land on Tuesday, or I would start looking at other options. He was not the only game in town.

When the next Tuesday rolled around, I got to the cafe at a time I knew his audience would be gone, and he would be headed to his truck. I intercepted him there. He seemed in a much better mood and not near as surly as before. I offered my hand for a shake and indicated I was ready to see that land. He invited me into his truck to go have a look. I crawled in and pushed his giant wok of an ashtray off the passenger seat and onto the middle console and closed the door. As we headed west down FM 11 to the property, he smoked nearly one cigarette every mile. I never saw anyone smoke so many cigarettes so fast.

The property was not much to look at—mostly mesquite trees and saltbush—but that is what most of that part of West Texas looks like. A deeper look is needed to find the potential of a piece of land in that area. I took samples of the soils and put them into plastic baggies. I asked Carl for permission to hire a backhoe and dig around to see what the soils looked like under the surface. He agreed to let me do that. The property had some old gravel pits that would be a good place to run the water to when it was time to drain ponds for harvesting. With frontage on Farm Road 11 and some electrical transmission lines running right down the fence line, it looked like all the basics were in place for this little piece of West Texas to become the first real, commercial shrimp operation away from the coastline ever built. As we drove back to the cafe, I told Carl that it seemed like this land would work and asked him to come up with a price. At that time, we still had that first crop to produce on the demonstration farm, and I needed to see good numbers before there would be an offer made, but I told him when the time came and if the price was reasonable, he would be the first person I called. He reminded me that I knew where to find him when I was ready. I don't think he cared about the shrimp farming idea; he just was happy that someone was interested in a piece of desert scrub land that he had no use for. As soon as the doctor gave the go-ahead, I went back to Imperial, headed straight for the cafe, and told old Carl we were all set to go. I had the title in my hands in just a few days and we were almost ready to start. The next important task was to arrange a place for my family to live.

By early July, in the unbearable heat of West Texas, I arranged to have a nice new double-wide trailer house installed on the property. I then called Ronnie to start the land clearing and get the whole thing underway. As the land was being cleared, I had some time to think about the engineering on this new farm and I started to have reservations about the quality of the pond construction knowledge

and skills available locally. Ronnie was a good equipment operator
and had a first-rate earth-moving company but, through no fault
of his own, his knowledge of shrimp ponds was limited to what
they had built on the water district farm. The water district ponds
were constructed just like oil field pits, with a flat bottom and thin,
very steep levee sides. That's all the local folks knew how to build
at the time. In the oil field, the pits just hold water, oil, or mud.
Trucks don't drive on the levees and the pits don't have to drain
clean and efficient, if they even drain at all. For shrimp farming,
ponds like this would not do. Flat bottoms leave behind shrimp in
the mud. Steep levees weather and erode quickly, giving the pond
a short lifespan and a lot of maintenance costs, not to mention
making driving trucks and shrimp feeders along the tops of these
weathered levees a perilous adventure at best. I knew I needed real
aquaculture ponds. It was time to call someone who knew how to
build ponds constructed for this exact purpose—it was time to call
in the experts.

OF DIRT AND WATER WELLS

Most of the shrimp farms along the middle Texas coast were built
by a group of Mennonites who lived in the Victoria/El Campo
area. They also built a few catfish farms in the same area a bit more
inland. Hector had a contact in El Campo I could call and find
out if there was a way to get any of those experienced shrimp farm
builders to come out west and work for us. This contact was indeed
a Mennonite with an earth-moving company and lots of equipment, but he said he had a nice, established business in that area and
was not interested in a ludicrous proposition like the one I had in
mind. Fortunately, he gave me the name of a couple of brothers who
were young, both newly married, and wanted to step out on their
own with their own earth-moving company. They owned the rigs
and the laser equipment necessary to make ponds with compound

bottom slopes and wide levees, but the equipment was located six hundred miles away. We spoke on the phone and once again, just like with the doctor and his son, I got someone I hardly knew all excited and fired up on the idea of shrimp farming in West Texas. I think they could sense my optimism, and with their youthful energy and general intrigue for something new and different, they said they were in.

In short order those young men got into their respective, giant articulated 4x4 International tractors, each pulling an enormous, laser-equipped scrapper behind it. They motored the backroads of Texas at 20 mph all the way from El Campo to Imperial, more than six hundred miles, while their wives followed behind in two pickup trucks loaded with all their worldly possessions. This traveling party arrived in their machines early one morning just at daybreak, and it was a remarkable sight. Those 12-foot-tall, red articulated tractors with equally enormous scrapers behind them pulled into Imperial looking like something out of a science fiction movie alien battle scene. While these machines were common in big farming country and many other places, no one in Imperial knew they existed. You would have thought a flying saucer had just landed in Pecos County. I was excited because these boys had wonderful vocational skills and understood earth moving and production pond building very well. They had made mechanical and computer modifications to the equipment so that it would do very detailed and intricate work. They also beefed up their scrapers so they would do more work and last longer. They were indeed talented engineers and just what we needed for this grand new adventure.

The folks in Imperial were dumbfounded and even more so when out of the machines came aliens of another sort: Mennonites. I think a bus full of traveling nudists would have gotten fewer stares and mutters. Now it is not that they looked so different. The two men wore jeans and work shirts and sported beards, while their wives wore plain dresses and had their hair up and covered. Few people from Imperial had ever seen a Mennonite, and their arrival sparked more than a few whispers. As for me, I was ecstatic to welcome these wonderful folks and to get them moving dirt. They were just good, clean, salt-of-the-earth Christians. Our families would become very close friends and spend a lot of time together during the next few years. These men ended up building every shrimp farm that operated in West Texas with the exception of the water district farm.

The first thing we did was to see the property where the farm would be located. By this time Ronnie had cleared all the mesquite and saltbush and the space was just an empty piece of land with red clay dirt—a clean new canvas on which to paint our masterpiece.

Earth-moving construction is an arduous life requiring tremendous skill, and these guys were better engineers than most of the engineers I met who had college degrees. These two fellows, Kelly and his brother Terry, were cream of the crop. They could design and build the most amazing earth-moving projects and mechanical projects, too. I had the biological and aquaculture knowledge, but they brought to the venture fresh eyes and the rigorous math and engineering skills that we would need for the operation to become a real industry. Proper equipment, proper pond construction, and a lot of fun and laughs, to boot.

Together we planned the twenty-five-acre farm. We took it very seriously as we knew that this would be a real challenge and there was so much money on the line. Even though it was not my money, I treated it as though it was, since my reputation was on the line as well. We spent many hours at my kitchen table platting the ponds. This was an important step, and the whole reason I hired these guys to come all the way to Imperial. I did not want small, flat-bottomed ponds with narrow, easily erodible levees like on the water district project. For this farm we would design two-acre, rectangular ponds with bottoms that had a compound slope that would drain to the middle and to the end simultaneously. This would get us a much better and "clean" harvest with minimal shrimp stranding in the mud at harvest. We had room on this property for fourteen of the rectangular ponds and one strange-shaped pond that had to be made in sort of a triangle shape just to fit the available space. We also blueprinted out all the ditches, water supply lines, and pond drain lines. We designed plans for the greenhouse and barn locations, where to run electricity, and where to drill the enormously important water wells.

We then surveyed the elevations across the property so we could calculate the yards of dirt to be moved and where it would be moved to. This is an amazing thing to see. Once we knew the ground elevations and nature of the dirt on the cleared ground, we input that data along with the pond boundaries into the tractor's computer which tells the machines where to pick up and where to put dirt down in a way that makes the most economical use of the land and the equipment. We marked the pond boundaries with little survey flags in the dirt and marked where all the lines and electrical were to be installed. These flags were the map that the dirt movers would follow when they started moving dirt. After a few weeks of planning, we were ready to start moving dirt. What a day it was to finally get moving in a tangible way. The big articulated tractors and scrapers can move so fast it is unreal.

Individual ponds were not built one by one—the whole farm took shape all at once. The lasers tell the tractor computer what the ground elevations are, then that computer program uses that information to tell the scraper when to cut into the ground and when to fill. Depending on the plan and the elevation requirements, the machines may relocate soil. The idea is to carry the dirt to where the plan says it needs to be by traveling the least distance possible. As the pond bottom is cut out, the dirt is used to make the nearest levee or berm. These tractors can also turn on a dime and they can cut so fine they could pick a quarter up off the ground and into the scraper with just a handful of dirt. They could also dig a full scraper load, move it fast, and disperse the dirt in exact measure anywhere across the farm. It was a sight to see. In just a few hours on the first day you could see the basic shapes of all the ponds and get a feel for what the whole farm would look like. No one in Imperial—or all of Pecos County for that matter—had ever seen such a display, and trucks and cars were always stopping on the side of the highway to watch the show.

Once the main body of the ponds took shape, and the levee slopes began to build up, the tractor operators started working on the pond bottoms. The dirt cut out of the bottoms was used to finish the pond sides and levees. The bottom work is the most delicate, most important dirt work of all. The lasers can be set to make nice slopes, even compound slopes, across the bottom of the pond. We designed the pond bottoms so they would slope toward the drain end, but at the same time, each half of the bottom sloped into the middle. This way the water in the ponds flowed to the middle, then down to the drain. This design enabled the water to flow swiftly and the shrimp to flow with it. It must be done correctly, or the shrimp become stranded and lie on the bottom in puddles, ruining in the heat while the pond drains. The plans we made for all this were very detailed, so the expectation was that these ponds were going to harvest clean with minimal stranding of shrimp. Unfortunately, we did not have a budget for a shrimp pump harvesting system. We decided to use the system identical to the water district farm, where the water was drained into the live car net affixed to the drainpipe and then positioned in the ditch on the back side of the drain levee. This was a labor intensive and cumbersome harvest method, but at least there would not be problems on the pond side of the drainpipe. The little farm was shaping up nicely, and I was indeed proud of how the operation was starting out.

Dirt work and pond construction begin quickly, but the finishing touches consume more time. Part of the pond construction involved the dirt work on the ditches that carry the harvest water away from the production ponds and into gravel pits. As Kelly and Terry finished the last of the pond work, the time came to start another major project that was every bit as important—the construction of a greenhouse to head start the larval shrimp. Unlike what was used at the demonstration farm, we decided to purchase a real, commercial greenhouse. Head starting would get us bigger shrimp sooner or possibly two crops a year. A real, professionally manufactured and constructed greenhouse could house good, sturdy raceways and control the temperature even in the coldest weather, allowing us to start the baby shrimp in the late winter without risk. We purchased the materials for a 30,000-square-foot Nexus greenhouse, double-dipped galvanized frame made with 4-inch-square tubing. It was the Cadillac of greenhouse structures. Too often people try to save money by purchasing a cheaper greenhouse only to end up spending more to repair or replace the cheap materials. They also may risk losing their crop due to the failure of lower quality equipment or materials, and that can cost them their whole business.

The buildout of this structure involved hiring a contractor approved by Nexus, the company that made the building, to install their greenhouses. This was a serious structure and not something individuals should attempt. On their suggestion, we hired a contractor from Michigan who had installed many of these greenhouses. In October 1993, the contractor showed up in Imperial in a Volkswagen bus accompanied by his hand, or helper. In West Texas and the Southwest, we call helpers, or workers, hands. You will often hear someone say to a helper, "Go make a hand," or hear a father ask his son, "Did you make a hand today?" That means you worked hard, or you were helpful and useful. Actually, to a good worker it is a valued compliment to have your boss introduce you as his "hand" or his "best hand." If he refers to you as just his employee, then you probably won't be around for long.

These two guys had never been this far south and the whole desert was a fascinating thing for them. What was also fascinating for them (but not in a good way) was the wind. It sucked. These guys needed to assemble a structure that was about 175 feet long by 20 feet tall which had to be bolted together after each component was erected. In a steady 30 mph wind, this was a nightmare. These two were tenacious and worked hard and steady all day long, every day. Slowly but surely the structure started taking shape and looking like a greenhouse. They slept at the farm in their Volkswagen bus and worked twelve-hour days, including weekends. We let them come shower in the house, and occasionally they would have dinner with us. My little son was intrigued by their "hippy bus" and asked to go see it every day. He did not have the concept that it was their home, but we would visit, and I would not let him impose on them too long. On October 31, 1993, we got a big snowfall. It was 85° Fahrenheit the day before, and we woke up to three inches of snow on the ground that morning. Those two guys were astonished. They carried on as if they had witnessed the Transfiguration. None of us could work for all the snow, so we just hung out and drank beer all day and told lies and stories and they talked about what a weird project and place Imperial was. By the end of the next day all the snow melted, and it was 80° again. Welcome to West Texas. Those fellows soon had the whole building finished sans the roof coverings.

Building the structure was one thing, but installing the roofs was another matter altogether. The roofs were a double layer of clear poly material like heavy cellophane: big giant sheets of clear plastic 30 feet wide and 175 feet long. Once installed they are pumped full of air between the layers and that makes a very sturdy and insulated clear roof. The problem with this type of roof is that the panels must be installed in calm, non-windy, not even breezy conditions—dead calm. These conditions don't exist in West Texas. West Texas is most famous for two things—oil and wind! It would also take all of us—me, Little Manuel, the contractor, and his hand—to install these huge sheets of plastic up on the roof of the greenhouse structure. We had to wait for many days and nights until we finally caught some calm conditions, and these conditions came at strange hours, such as 3 o'clock in the wee morning. We had to move fast because once we were committed to the huge sheet of plastic, we had to get it up on the roof and secured. If any breeze picked up, the plastic became a giant sail . . . you can guess the outcome, especially if you felt the need to hang on to the plastic. There were some close calls, but during a two-week period we got the roofs on and secured. Once the greenhouse was erected and the roof installed, we had a climate-controlled environment in which to work. We could begin work on the raceway tanks on the inside, and weather would not be able to delay us or even make it uncomfortable.

We built eight large tanks, 150 feet long by 10 feet wide, inside the greenhouse building. Each raceway tank consisted of concrete foundations with 1-inch plywood sides and then lined with a thick plastic material on the inside to hold the water. Each of these raceways was a thing of beauty for sure. Next, we built a nice drainage system that channeled the water to a harvest area outside of the building where we could collect the juvenile shrimp and transfer them to the waiting ponds once the weather was warm and suitable. We completed the electrical systems and water lines inside the building—we just needed baby shrimp.

The pond dirt work and the greenhouse construction were completed at about the same time in early November 1993. It was time to have water wells drilled and big PVC water lines to the ponds installed. I called my friends at West Texas Water Well Service whom I had met on my first day in West Texas, and they went right to work drilling three wells with a modern rotary rig that drilled very fast. It took longer to rig up and rig down than it took to drill the wells. When we started, we had no idea how deep the wells would be. The old timers in town told me that once we hit the red bed, a red layer of clay hundreds of feet thick, we would be done because there was no water below that. I spent a long time with a local old man named Pete Yodor. He was the board president of PCWID No. 2 and had lived in the Imperial area his whole life. Pete had done it all, from farming to drilling oil wells to running the water districts. He was a very kind and helpful man and provided me with lots of advice and guidance about drilling water wells in the area. We ended up hitting the red bed at about a hundred feet. The water table started at twenty-eight feet, so there was seventy-two feet of water. That is not a lot of space to pull water from at high volumes, but a 14-inch casing was set with the plan to use 10-inch pipe. A 10-inch well pipe with the right horsepower pump will carry a lot of water. Most fortuitous was that this formation was very porous. It was a water bearing formation called the Pecos Alluvial saltwater aquifer, often referred to as the Permian Sea. This formation was made of large gravel, so the water just poured into the well casing, replacing the water just pumped out in milliseconds. Even the well drillers had never seen anything like it before. These wells could extract 2,000 to 2,500 gallons per minute each. We could put that underground ocean onto the surface fast.

With the three wells drilled and pumping, the ponds built, and the greenhouse ready, all we had left was to plumb it all up so we could get water everywhere we wanted it. We had strategically designed the 12-inch water line system so that every well was plumbed into the system to maintain good pressure throughout the farm. There was a water inlet to every pond and a water line to the greenhouse providing water to all the tanks inside. Drains were located at one end of every pond and were how we would get the water and the shrimp out of the ponds. Drains are tricky—you must dig through those beautifully finished levees to set the pipe. This is a challenge for even a good backhoe operator, and we didn't have one of those. When we started the project, our contractor Ronnie had a hand that was a top-notch backhoe operator. My three-year-old son named him Backhoe Ray. He was sometimes referred to as Crazy Ray by local folks, which was also accurate, but around our place we stuck with the Backhoe Ray name out of respect. He operated the backhoe boom and bucket like it was his own arm and hand.

Unfortunately, just after we started construction, he and Ronnie got into an argument, and he quit. Backhoe Ray would eventually buy his own backhoe to do work for hire, but in the meantime, we had to hire the only other available backhoe operator in town. I think the first day he worked on the farm was the first time this guy had ever even seen a backhoe, let alone operate one. Needless to say, the job took quite a bit longer than we had expected. With time, this guy got much better as an operator, and the job eventually got done. He dug the ditches, Little Manuel and I put the pipe together and set it in place, and then he covered it all up with dirt using the backhoe. In a few weeks we were plumbed up and water was flowing. We also installed the electrical lines with the plumbing, so everything was finished in early January 1994. I did have one little run-in with the electric contractor and an interesting lesson in how business was often done out here in oil field country.

We hired an electrical contractor out of Monahans, Texas, located twenty-nine miles up the road, to build the electrical system on the farm. These electrical workers showed up with a Ditch Witch machine to cut a few ditches to lay electrical conduit in and bury all of it. The first day they showed up, they made about 20 feet of ditch (maybe thirty minutes of work) and then the ditching machine died on them. They could not seem to get it to start again so I decided to see if I could help. I fancy myself a fair mechanic and figured I could be of some use. I ended up spending about five hours getting that machine to start up again for them. By the time it was finally able to run and dig again, it was time for the contractors to go home for the day. They came back the next day and got their ditches dug and lines laid. In a couple of days, we got their bill, and those pinheads charged us for a whole day that first day! Charged for the machine, two men and a truck. I was livid. The secretary told me that they usually do oil field work, all the oil field offices are in faraway places, and these out-of-state oil companies just pay the invoices and never question anything. I don't have a lot of sympathy for oil companies in the first place, but this was just thievery in my opinion. I let the electrical contractor know that farmers can't afford to give away money like the oil companies can and I would be scrutinizing their invoices from then on. It turned out to be good practice because in the years that followed I would find this sort of sneaky billing more than once by numerous companies. I told myself I hoped I would never get so big or prosperous like the oil companies that I would just allow people to rip me off, and well, I never did.

By spring 1994, the farm was operational and shrimp production was underway. Learning to operate a new farm was a lot of work and especially in such an odd place to be growing shrimp. Manuel and I had a lot on our plates. Help was almost always appreciated. That summer my wife's fifteen-year-old nephew Jeff and his friend came to work for me on the farm while they were off on summer vacation from school. It was just Little Manuel and me working on the farm, and I felt that we could use the help, especially free help. After all, I had spent some amount of time every summer from childhood to college age on my grandparents' ranch near the Red River in North Texas. That place was my nirvana. I spent most of my time fishing the stock ponds for bass and catfish, but I made a hand with the cattle when called to do so. I felt my nephew could provide good help and make himself some lifelong memories as well. Not everyone gets a chance to go to the desert and work on a shrimp farm. Actually, at that time, no kid or adult had ever had the chance to work a shrimp farm in West Texas since it had never been done before! It was unusual enough that when he went back home to the DFW area and had to write the requisite "what I did on my summer vacation" essay for school, his paper describing his summer on a shrimp farm in West Texas got him scolded and sent to detention, accused by his teacher of making the whole thing up!

As it turned out, Jeff and his buddy were pretty good hands on most days, but you would have thought they had never been out of the sight of their parents before, and maybe they hadn't. They worked hard in the daytime and catted around town all night. It was all I could do to keep them out of trouble. One night they took the truck and drove it at full speed up and down every oil field road in Pecos County. When they were not doing doughnuts in the truck, they were shooting jackrabbits with whatever gun they had managed to sneak out of the house. One day on the farm they managed to get the little feeder machine stuck inside a pond to the point we had to get a tractor to pull it out. I told them that they were supposed to put shrimp feed in the ponds, not the feeding equipment. It was not long after that those two fifteen-year-olds went to town one night, found Little Manuel, and got him to buy them some alcohol. Keep in mind, Little Manuel was only about nineteen at the time and like most young folks, was always up for a good time. They all proceeded to get snot slinging drunk and stayed out until the wee hours of the morning. They were in terrible shape that next morning—the boys were still drunk and puking their guts up in the bushes out in front of the house.

My wife was furious, to say the least. She gave them the lecture of a lifetime, coming just short of wringing their necks. I thought it was kind of funny, maybe because I also knew that this was a workday, and I would have the opportunity to make them regret every sip of whiskey or whatever they had drunk from the night before. I paraded them out to the farm and put them to work. We had a load of shrimp feed in 50-pound bags that needed to be neatly stacked in the barn—40,000 pounds' worth. That was on top of the feeding and pond checks that were routine each day. I probably came up with plenty more tasks that day as well. They were miserable. They never asked to go to town or borrow the truck for the rest of the summer and became respectable citizens and farm hands from then on. There were a few other dramas that summer, like the rattlesnake that went straight for Jeff's private parts for some reason. He narrowly escaped with his manhood, but sometime later he did chop the end of his finger off with a very sharp knife that, according to Jeff, "I just found in your room and was only looking at." They also managed to run over and destroy a borrowed generator that my friend had lent to me to use. Of course, it was my buddy's most "prized possession," so I had to pay a hefty price. Overall, I hope the cherished memories those boys made that summer were good ones because they cost me plenty.

The operational plan for production was a double crop of shrimp that would be head started in the nice new greenhouse. We would introduce larval shrimp into the greenhouse in February, move the shrimp from the indoor raceways to the outdoor ponds in May, and then harvest from the outdoor ponds in early July. After we moved the juvenile shrimp from the raceways in May, we would immediately introduce another batch of larval shrimp into the indoor raceways. After the July harvest of medium-size shrimp from the outdoor ponds, the second batch of greenhouse shrimp would be moved into the outdoor ponds, and we would get a second harvest of medium-size shrimp by late October. This was a very nice plan, and getting two crops of medium-size shrimp was a dream many shrimp farmers on the coast had wanted to do for years, but with limited success. We perfected it the first time we tried it. The only real problem was that medium-size shrimp stopped bringing a good price shortly after this season. When the price for medium-size shrimp, which is a commonly used shrimp in the food service industry, was solid, the market would become flooded with domestic shrimp from the wild-caught industry. Medium shrimp was and is still the main shrimp the inshore and bay shrimpers catch in their nets, so before long the market was oversupplied, and these sizes were not valuable enough any longer to try to grow on a shrimp farm. The beautiful thing about a nice greenhouse is that you can use it to either produce that two-crop scenario, or you can head start the larval shrimp in the late winter/early spring and then put the shrimp in outside ponds and grow all summer. This extends the growing season and gets you good-sized shrimp that are always more valuable in the market. This would be the last time anyone produced a two-crop system in West Texas. The larger sizes just became worth so much more that we focused on producing as many large shrimp as we could in all the future seasons.

Now that we were in production, the Triton Aquaculture company, as it was named, needed to devise a plan on how to process, freeze, store, and sell this unique product. We found a processing plant in Hobbs, New Mexico, located about three hours away from Imperial. Three hours was a short distance in this part of the world and practically in your backyard. In this part of the country, we often use time rather than miles to determine how far things are since everything is far away; using time as a reference helps you plan logistics much better than using mileage. This processing plant had been in business for years, processing all sorts of products from meat patties to onion rings, though it was struggling to stay in business. What interested us along with the proximity to Imperial was that they had a beautiful spiral cryogenic freezer. A spiral freezer looks like a spiral staircase but instead of steps going up in a circle, it is a metal mesh conveyor. The raw product is placed on the spiral at the bottom, and as the spiral turns the product moves up the conveyor and is individually frozen very quickly. At the top, a spray bar coats the product with mostly water, producing a frozen layer that protects it from freezer burn. The individual frozen shrimp then fall off the top of the spiral conveyor into waxed boxes which are stored in a freezer. It's a sight to behold. These folks had never processed shrimp or any seafood before and it took several weeks to get the approval from their USDA (United States Department of Agriculture) inspectors to process seafood, but once they did, we gave them a crash course in shrimp processing. Shrimp were usually processed by grading for size, de-heading, and then freezing. However, freezing the shrimp with their heads intact was becoming more popular, and since that was a much cheaper method, we went with that method. We also decided not to grade the shrimp since they were all reasonably close to the same size.

THE LOST SEA–WEST TEXAS DESERT SHRIMP BRAND

We had our first shrimp harvests on this farm in 1994. We made a simple and straightforward harvest plan. The farm had fifteen ponds, and each time we harvested a pond we would send that pond to the plant to be processed. We had lined up a fellow with a truck and trailer to haul the shrimp each harvest morning to Hobbs. The only weak link in the whole system was that ice was so hard to get and keep in West Texas, especially in the quantities needed. We struggled with having enough ice to keep the shrimp from spoiling in the hot weather. The first load we sent to Hobbs was completely rotten by the time it got to the factory, only four hours later. The hands did not have enough ice to chill the shrimp properly and they did not put water in the totes with the ice and shrimp. It is critical to add some water to the ice, as the ice cooled the water, and the water got in between the hot shrimp and cooled them down as well. It works best when you also stir the tote of shrimp, ice, and water to really cool everything down. Otherwise, the ice will preserve the shrimp on the outside, but in the middle of that 1,500-pound tote, the hot shrimp will just start rotting on the inside right from the start. That is exactly what happened with the first shipment, and I took the heat for it and rightfully so as it was indeed my fault. I should have given better direction to the harvest hands and been more vigilant as to how the totes of shrimp were handled during the harvest. The doctor chewed me out for this expensive mistake, so I set about to make it clear to everyone how to ice down harvested shrimp the correct way. From that point on, all the other ponds made it to the processing plant in perfect shape.

Earlier that summer, soon after we had found the Hobbs processing plant, I started working on a packaging and marketing plan with the doctor. We also built a large walk-in freezer on the farm that would hold most of the shrimp we could produce. We finally agreed to have the Hobbs plant individually quick freeze (IQF) all the shrimp with their heads on and package them in 2-pound bags with our brand name and logo printed on the front. We would store them in our freezer until we could sell them. The only problem was we didn't yet have a logo or a brand name for our shrimp. To be more specific, we didn't have a brand that the doctor, his wife, and his son had approved. I had come up with what I thought was a cool name: Lost Sea–West Texas Desert Shrimp. I figured since we were sitting on top of the old Permian Sea and using that water for our ponds then we should capitalize on that idea. It was unique and caught your attention and imagination. Doc and family wouldn't have any of it and felt that we should hire an expensive advertising and marketing firm and let them come up with a name.

The doc's wife found a Madison Avenue advertising and marketing firm and sent them my suggestion of Lost Sea and a few more that she had thought of. After about a month, the advertising firm came back to us with a big invoice and the very best and most marketable name and logo that all their research and focus groups had suggested: Lost Sea West Texas Desert Shrimp. Their suggested logo also came from what was our current company logo, which was a drawing of Triton, god of the sea, blowing his conch shell with his trident in hand. This had been our company logo since we first started up back in Florida, but the marketing firm suggested redoing the artwork to give it a more modern look. Their suggestion looked fair at best. One of my friends told me that instead of looking like the Greek god Triton blowing a conch shell, it looked like a homeless guy without a shirt drinking liquor from a paper bag, but I didn't protest to the doctor and crew. I actually felt gratified that I had come up with the brand name, and I could live with the odd-looking, vagabond god and his paper-bag-looking conch shell.

We didn't set the world on fire with our production numbers, but we did produce some reasonable crops of shrimp that year. The presentation of the product was very eye-catching, with the beautiful, individual quick-frozen shrimp visible inside the clear bags with the Lost Sea name printed in yellow across the front. We even had a description about the Permian Sea and the clean water, and all printed on the packaging so buyers would know this was not something from the Gulf of Mexico or South America. This was shrimp from the Texas desert.

The Hobbs processing plant did a nice job considering their inexperience with shrimp and really did us proud. I wish that we had been able to use their services and that nice plant for years to come but they were on the brink of bankruptcy when we found them. Working with our product kept them afloat for that summer, but it was not enough to keep the place from going under. By the next fall they had shut the doors for good. But what a nice spiral freezer! I always wondered what ever happened to that beautiful specimen of modern processing technology.

The rest of that fall and into winter 1995, the shrimp sales were brisk, prices were high, and the margins were good. We sold mostly to a couple of grocery stores in Odessa and Midland and to the local public. The one-farm shrimp industry in West Texas was off to a great start. I, however, was not doing so great. It was becoming incredibly hard to work for the doctor, and his son and the son's wife were even worse. None of them knew much about agriculture and absolutely nothing about shrimp farming, but you could never tell them that. They thought they knew all the answers, or at least expected others to treat them as if they knew everything. It was also becoming apparent that this whole business was set up to give the son gainful employment. Soon the son was doing all the planning and giving all the orders from his home in Florida—he did not want to move to the desert. After all, he was a doctor's son with a wife who was a socialite who had airs to put on, and they were all super proud of themselves and their status. They came to the farm from time to time but did not stay long, creating a very stressful environment when they showed up. They would start with questions meant to make it clear that our ideas were all wrong and we were not doing enough to make things work to their satisfaction. Then they moved on to belittling everyone and everything in sight.

When they were satisfied that they had put everyone in their place and had made clear the class distinctions, they patted us on the back, wished us well, and left. Little Manuel so hated it when they came from Florida to the farm that he would find any way he could to disappear as soon as they arrived. His distaste for them was so great that once, when what was an obvious rental car pulled into the drive and up to the greenhouse door, he took one look at the occupants of that car and made a beeline out the back of the greenhouse and I didn't see him the rest of the day. The next day I inquired of him what made him leave in such a rush. He said when he saw the son and his wife drive in that he was out of there as fast as he could fly. I burst out laughing and told him that what he saw yesterday in that car that drove up was not the son and his big-haired wife but a feed salesman with a big white Standard Poodle sitting next to him! We got a laugh out of that we never forgot.

In all seriousness, I was heartbroken by the way things were turning out at Triton Aquaculture. I put everything I could into that farm. It was the first purposely designed and engineered commercial inland shrimp farm anywhere in the country. I really loved that farm, and I always will. My second child was born on that farm, and we had a lot of sentiment attached to that piece of West Texas desert scrubland, but I knew that I had to do something else because the situation was not going to work for me. I wanted to see an inland shrimp farming industry become a reality, and I did not intend to be the pool boy for some rich, spoiled punk.

That winter, I gave them notice that I would be leaving before the second growing season, but they seemed to think I had some legal obligation to stay there and be the target of their aggressions and insults. I made it very clear that I had no intention of staying for any length of time, but I would help them transition to another farm manager. I stayed long enough into the next spring season to help them start the larval shrimp in the greenhouse. They could not find anyone who would come to West Texas and work for them, so the son was forced to come to Imperial and run the farm after I left. He put on a brave face, but he hated every minute in West Texas. I can only imagine how miserable his wife was too. He did, however, make a reasonable crop that next summer with the shrimp

I helped him head start in that awesome greenhouse. As soon as he harvested that crop and loaded it in the last truck for processing, he packed their bags and went straight back to Florida. Sadly, that farm sat idle ever since. They sold it to a guy from Midland who called it Pecos River Aquaculture, but all he ever did was rent out a couple of ponds to someone who tried to grow a few shrimp and some striped bass, most of which died from neglect. It was a real shame. It now sits sad and alone with mesquite trees and salt cedars growing up through the bottom of the ponds. That amazing greenhouse is now nothing but the skeletal remnants of those beautiful silver galvanized metal frames. They stand proud in the desert sun like a giant tombstone marking the grave of a once noble effort.

THE IMPERIAL SHRIMP COMPANY

Soon after Triton Aquaculture broke ground on that first commercial farm, PCWID No. 3 mothballed the demonstration farm. Hector, Big Manuel, and Danny were let go, and Rosie retired altogether. I guess the district felt they accomplished what they set out to do, which was attract a commercial venture, and with that, their job was done. They really did not want to run a shrimp farm, nor did they want to bankroll the expenses of the facility just for university researchers and students to do their experiments. Their goal had been to prove it was feasible, attract private investment, and then sit back and see what happened.

I was urgently trying to find a way to continue to shrimp farm in West Texas, but not for Triton Aquaculture. Fortunately for me, during my time in the area I had met a local Imperial fellow named Paul who had been on the water district board of directors when they first started the demonstration farm. He had declined to run for election to the board when his term ended, but he was still very interested in shrimp farming. He was not a farmer or agriculture guy at all, but felt shrimp farming might be profitable in numerous ways. He was mostly interested in selling shrimp. That first season that the water district project produced shrimp, Paul made a deal with Hector and me at harvest time to purchase a sizable quantity of shrimp, which he intended to sell all around West Texas. He outfitted an old semi-trailer as a processing room behind his house, and he and his family de-headed the shrimp, bagged it up, and sold it fresh. It was a lively little side hustle for him, and he and his family seemed to enjoy doing it. One day I ran into him, and he mentioned that he wanted to rent the now idle demonstration farm from the water district, get it up and running, and grow shrimp there. He asked if I would be interested in helping him in this venture. While I had a feeling he did not really know what he was getting into, this was music to my ears. I knew that little farm with all its design problems would not support even one of our families, let alone both, but it was still something that might give me the opportunity to change course.

Paul and I made a plan, then took it to the district to see if they would lease us the farm. They agreed to lease it to us for the next year. Next, we took our plan and visited every bank in the area to see if anyone would loan us a little money to operate the project. I was not ready for the initial rejections we got, but then I was young and green at that sort of thing. First, this part of West Texas is no farming mecca. Cattle ranching is the agriculture of choice and an enterprise usually done by folks who have lots of oil wells on their land. They say three cows per pumpjack is the perfect stocking density. The banks don't easily lend money for agriculture in that area, and top that with a type of agriculture no one had ever heard of and scarcely believed even existed and the result was that we got some very painful comments from bankers. I remember one banker said, "Y'all are probably going to get rich, and then I will regret that I turned you down." I told him that's what the record executive who turned down the Beatles said and that he should want to learn from history. He said, "You can just come back in a few years and tell me what a stupid jerk I was for not loaning you the money," and I told him I would just tell him that now and save myself the trip back.

We went back to Imperial and commenced to upgrading the demonstration farm as much as possible with the limited funds we possessed. I had virtually nothing but experience and knowledge, but Paul and his family had some money, and they were willing to invest a little. Paul's father was apparently a successful El Paso businessman and agreed to give us a small loan. We could not redo the whole place with the small loan, but we were able to build a better greenhouse than those simple greenhouse coverings that had been used on the demonstration farm, and we purchased some equipment that would make the shrimp farming there a little easier. Mostly we just had enough money to put in a crop for that next summer. Feed, seed, electricity, and a little labor.

We head started the larval shrimp in the greenhouse in March and then put them outside in the ponds as soon as the weather warmed in May. We hired Big Manuel to feed the shrimp and take care of the farm because he knew the farm well and had experience shrimp farming—and he worked cheap. Paul had a regular job at a gas plant in the area and I was still with Triton for the meantime, so neither of us could be at the farm much at all. I stopped by every day in the afternoons after taking care of business at the Triton Aquaculture farm. The two farms were adjacent to each other, so it was really just like working on a big farm with more ponds. Without much fanfare, Paul and I managed to produce a crop of shrimp on that rented farm that summer. Paul and his family processed it and sold it through the channels that they had developed selling the water districts shrimp a couple years before. It was the Imperial Shrimp Company's shrimp. It was a good business, but unfortunately it was not profitable enough for us to continue. We just did not produce enough that year. That farm was not well suited to commercial production and was inefficient to operate; more important, you can't run a shrimp farm part time. It takes constant supervision and long hours to do it right. It was really something we, and especially I, should have known better not to do. I wanted to do anything but continue working for the doctor and his son, so my judgment wasn't very good. Paul loved this little shrimp-selling business, but in reality, he was twenty years into his gas plant career, and it made more sense for him to stay focused on that and the long-term security it offered him and his family. The Imperial Shrimp Company venture did not work out beyond its first year, but what it did do was help continue the industry and put more of that awesome shrimp out there in the local market. Because he sold the shrimp everywhere and to anyone, it was really Paul and the Imperial Shrimp Company that started to get the attention of the West Texas region. Because of Paul's and his family's efforts, everyone in this part of Texas became a fan of that sweet, delicious shrimp from Imperial.

CHAPTER 5

THE SHRIMP FEST

T o top things off, during these early years of the shrimp industry we had a genuine Texas fandango, the West Texas Desert Shrimp Festival. Actually, we had two of them. That first year, 1994, was the year that we hatched the idea for a festival of some kind to celebrate the shrimp farming ventures. Since it looked like we would for sure make a decent crop of shrimp that year, Triton Aquaculture farm, with help from both water districts and the then Pecos River Compact commissioner Brad Newton, decided to have a great big party to draw attention to the new industry and introduce folks to the Lost Sea–West Texas Desert Shrimp. We wanted to show off our primo product, PCWID No. 3 needed to show the locals all that money spent on the shrimp farm demonstration project had not been wasted, and Brad wanted to show economic progress along the Pecos River in Texas. We chose to have the festival at the Imperial Reservoir. This is a small, 1,500-acre lake of impounded Pecos River water used to irrigate farms in the two Imperial water districts. Since there is very little farming and thus very little irrigation, the lake usually just stays full of water. Water Irrigation and Improvement District No. 2 actually owned the reservoir, and they would open it every summer to the public and charge a fee for people to swim or boat or just hang out.

CHAPTER 5

The Imperial Reservoir became the only real source of income for District No. 2 and practically the only summertime recreational space folks in the area could access.

The reservoir was a very popular and storied place. Fables from the reservoir fueled a plethora of gossip. According to the chatter, many important figures in town had allegedly participated in some illicit and usually immoral episodes originating from activities at the reservoir, tarnishing their otherwise good name, and the townsfolk would never let those escapades fade from memory. One story was tragic. One fine Saturday, when the reservoir was full of folks, a local fellow who was also a pilot chose to take his airplane and fly some acrobatics in the air above the reservoir to impress the onlookers. Unfortunately, he decided to do this air-show after being at a party of some kind and consuming way too much whiskey. He was in no shape to fly, and certainly not to show off. When he got into his plane, three sheets to the wind, what he didn't know was that a young kid had sneaked in with him to take a little ride on the sly. It could have been a relative or the child of a friend; I never actually learned who it was. As his air-show went on, the pilot got braver and braver until he was flying just above the water to the amazement of all his audience. During one of his daring, barnstorming maneuvers, the plane's wing clipped the water and the plane instantly crashed into the reservoir. The pilot somehow got himself out of the sinking airplane, but since he had no idea the kid had snuck into the airplane, he made no effort to look for him. The kid drowned, and the pilot went to prison. Fortunately, most of the stories were not as sad—a few rumors of wayward wives who took brief leaves of their domestic duties to shack up with drifters, or tales of men who went out there to settle a score, or score something illegal. For most people, the reservoir was just a place to go in the summertime to play in the salty water, and for a short while, pretend you were anywhere but the Chihuahuan

Desert. The Imperial Reservoir is not open to the public any longer and that is a real shame.

The Imperial Reservoir was a large area, and virtually the whole circumference was like a beach, but no one was picky as it was the only beach around. There were bathroom facilities, covered picnic tables, potable water, and garbage dumpsters. All these conveniences and amenities made it a suitable location to hold our shrimp festival. That first festival was a modest affair and not much turnout. I guess the whole thing sounded preposterous to West Texans not used to eating or being around any kind of seafood, let alone seafood grown in their own backyard. In addition, the district didn't really know what to expect and how to plan it. The first West Texas Desert Shrimp Festival wasn't a bust, but it was pretty dull. The second and final festival in August 1995, however, was a major success and drew thousands of people to the reservoir. By this time, there were three operating shrimp farms: Triton, Imperial Shrimp Company, and a new farm I had just gone to work for called Regal Farms. By this time, more West Texans either had purchased and eaten the shrimp or seen or heard of the shrimp farms in the Imperial area. In the past few months leading up to the second festival, I had done an interview for a CNN reporter, had a write-up in the Midland and Odessa newspapers, and Paul and his crew had been selling shrimp from Lubbock to El Paso. We were out there now, people knew about us, and they were very interested in this new and different sensation happening in the area. Something, anything different from the oil field and the same old same old. The folks were excited.

The second Shrimp Festival turned out to be the biggest party ever held in north Pecos County to this day. We had live bands and a shrimp/redfish cook-off. The cook-off was surprising to me. There were fourteen cooking teams competing in redfish and shrimp. The redfish, *Sciaenops ocellatus*, also called red drum, were some that Paul and I had inherited (still swimming) when we leased the water district farm. After that first shrimp run, they put in some baby redfish but started the process of closing the farm before the fish had grown up. They did keep them alive, and when we leased the farm from them, we got the pond of redfish by default. They were nice fish, but we could not give them away. No one in the area knew anything about redfish and would not buy them from us, so we harvested them and provided them to the cook-off teams at the festival. We thought this might generate some interest in redfish and help us get rid of them at the same time. The food that these cooking teams came up with was stunning. The very best part of that whole festival for me was judging that cook-off. I got to eat some excellent food for sure. You also get to drink a bellyful of beer because you are required to cleanse your palate each time you tried a new dish, and cleanse I did. Being a professional cook-off judge is a career I would definitely consider in my next life! That cook-off encouraged me about the possibilities the desert presented. If folks who hardly ever had access to good seafood, especially shrimp, could cook like this, then I figured they would buy lots of shrimp from us in the future. They would support the grocery stores and places that sold our shrimp. They would have backyard barbeques and cookouts with their families and friends, each trying hard to impress his pals. Everyone loves shrimp, and I guess I learned that just about everyone can cook it, too. They did a good job with the redfish as well, but that contest was mostly to see whose fry batter was the best. There wasn't much originality in the redfish dishes, and the teams that did take the redfish to a level above boiling oil were the ones who got the trophies—but boy, were they creative with the shrimp. In the end, the winning entry was shrimp-stuffed jalapeños wrapped in bacon. You can't lose with bacon.

Several Odessa/Midland radio stations covered the festival all day, broadcasting from the middle of all the action. Newspaper reporters were there, and all the local politicians came. The politicians needed to be seen supporting this promising new industry and give the impression in front of their constituents that they might have had a hand in its formation. The party lasted well past the 8 p.m. closing time listed on the brochures. Triton Aquaculture with its Lost Sea brand, Imperial Shrimp, and the newest farm in town, Regal Farms, all cooked shrimp and sold plates to the public. We wanted everyone to taste the shrimp and get into the whole idea of a desert shrimp farming industry.

The beer and seafood flowed all night, and all kinds of local, mostly bad, live bands played into the wee hours of the morning. Only one fight was reported, and that was a fight about musical tastes. This one and only fight was instigated by a young woman from Imperial. West Texas men are a rough and tough lot, but I must admit that a good number of the West Texas women are tough and opinionated. In this case, it was a preference for country music instead of other options. While we had invited several bands to play music that night, there was a seemingly disproportionate number of very loud, heavy music groups, head bangers if you will. As the night progressed and the alcohol flowed, most of the sounds from some of the bands became annoying to a few listeners. At one point, the local gal had enough and started requesting country music, adamantly. A guitar player from the metal band on stage at the time responded negatively to those requests for country music. Whatever he said or did infuriated her, and up on the stage she went. She proceeded to work him over pretty well. I was among those who thought the guitarist should learn to strum a few country tunes—just for personal safety and all.

The brief throw-down aside, the second Shrimp Festival was a pretty wholesome weekend all around and would not be soon forgotten. The best part was we got lots of press and publicity for the shrimp farms in the area and lots of folks got to eat the shrimp and learn about West Texas shrimp farming. Locals from Big Spring to Alpine learned all about us, believed in the idea of the shrimp farms, and were fans of the product after that weekend. For the industry to flourish, however, the shrimp grown in this perfect environment needed to find a larger audience. We needed the whole state of Texas to know about us, not just the western half. The industry would have to get bigger and more serious to make that happen.

One morning while I was taking care of some chores at the Triton farm, a car pulled up to the greenhouse and several folks got out. I went out to meet them as I figured someone was lost or something. Not many tourists came to Imperial, and it's not on the way to anywhere. If you're in Imperial, you either mean business or you're really lost. A very tall, fit, and serious looking man got out of the car and approached me. He asked if this was the shrimp farm he had heard about and if we could talk about the shrimp business for a bit. The other passengers, whom I assumed were his family, waited by the car. He told me his name was Dan and that he was interested in starting a shrimp farm in West Texas. I guess he had heard about us from the CNN news article that had been released, or perhaps he really had been studying the industry and read about what we were doing in some aquaculture trade papers. Our inland shrimp farming was starting to get the attention of the greater shrimp industry, even for just the novelty of it. That novelty never wore off, even after the shrimp production in West Texas achieved more than 500,000 pounds per year.

In any case, Dan was serious and was well schooled in the whole endeavor. He asked me lots of questions, some of which I could answer and many of which I couldn't, even though by this time I was pretty sure that I would be leaving the Triton farm for good. Dan told me he had a brother-in-law who was a mostly unsuccessful chili pepper farmer, but he thought he might have enough experience to manage a shrimp farm. He asked me if I would hire on as a consultant to help him and the brother-in-law get their farm started. I told him that unfortunately I would not be able to work as a consultant for them but that if he decided he needed real experience, then he could offer me the manager job, and I would consider accepting the position. I was thinking to myself, "What did I have to lose?" I wanted out of my current position and this guy seemed no worse than the folks I was currently working for, so I threw out the suggestion. He seemed very surprised but not in a bad way. More like in a "I was hoping you might say that" way.

Dan and his entourage eventually left, and I did not hear from them again for months. Just about the time Paul and I had started the Imperial Shrimp Company and were committed financially to that venture, Dan called me and said that the brother-in-law did not want to come to West Texas and would I consider working for him as manager of his, yet-to-be-built, shrimp farm. One of the other people in the car that day was the brother-in-law's wife. She apparently had a good tour of the Imperial area during that excursion and promptly decided, as most folks do, that she would have nothing to do with it. It looked like brother-in-law would stay a struggling chili pepper farmer for a while. I asked Dan to present me with an official proposal and I would consider it. Shortly thereafter, he presented his offer and job description and I accepted. I told him that I had to finish closing the Imperial Shrimp Co. and also stay on with the Triton Aquaculture farm long enough to get their 1995 crop started in the indoor greenhouse in February as I had already promised the doctor I would do. I could however work on the side to get his farm started with the goal of also having a crop that same year. Once I fulfilled my obligation to Triton, I would be all his full time. He was fine with that because he still had to finalize his land purchase and take care of loose ends. Even though all I could think was, "here we go again," I had better information and much more experience from my time in West Texas and I felt like I could do an improved job with a shrimp company this time around. I felt confident I could build a farm even better than the one I had built for the doctor and his family: one that would produce significantly more pounds of shrimp and make the shrimp industry take West Texas shrimp production seriously.

Once I had finished stocking the baby shrimp for the doctor's son and handed him the keys to the Triton Aquaculture shrimp kingdom, I packed up my trailer house and family and moved across Farm Road 11 to my new job at Regal Farms. I had called on my earth-moving Mennonite friends to start helping me design this new farm as soon as I knew I would be changing jobs, and dirt-moving work was already underway. The Triton farm had been about twenty-four acres, but this farm was going to be closer to fifty acres, so even with a couple months head start we were going to have to burn the midnight oil. It was February 1995, and the goal was to stock larval shrimp by that May. It was almost crazy to think we could meet our goal in such a short amount of time, but then again, the desert makes you crazy.

CHAPTER 6

REGAL FARMS

The property Dan had purchased had been an overgrazed cattle ranch with all the vegetation cut low to the ground with no trees or anything to have to clear off. We chose a spot on the 1,400-acre property where there were two large irrigation wells and close proximity to power lines. There might have been crops or some sort of cultivation grown at some time in the past in this particular spot, but it all looked more like pasture when we started. We never figured out why the wells had been drilled in this area, but it certainly was fortuitous for us. We could not have asked for a better starting place. With not much clearing to do, and a well at the north end and a well at the south end of the area, we could easily see in our minds how to lay out the ponds. Based on my experience at the previous two farms, I wanted the ponds to be a bit larger than I had built before. We designed them at 3.75 acres each. With reasonable production, each pond would produce enough shrimp to fill a refrigerated semi-trailer.

There was no processing plant available in West Texas anymore, but that did not matter for the new venture. Dan did not want to sell shrimp locally, except on a limited basis at harvest time. He wanted to grow many shrimp and sell them to the processors on the coast, just like a shrimp boat or coastal shrimp farm would do. Under this scenario, the majority of the shrimp would be trucked to the coast where facilities existed to handle the shrimp, process them, and store them in cold storage. Since we did not want to have so many harvested shrimp that we risked overloading a truck, nor did we want to send a truck off partially full, we figured a 3.75-acre pond was about the perfect size. Down on the coast most shrimp ponds were five acres in size and my shrimp farming friends down south were puzzled by the smaller ponds I had planned. I explained that their farms, located on the coast, were just minutes away from the processing plants; they could easily have a truck come back and pick up any excess shrimp that didn't make the first load. Imperial, however, was ten hours minimum from the coastal processor. We needed to stock a truck for each pond, send it away full, and only have a small amount left to sell locally. According to my best calculations and previous experience, 3.75 to 4.30 acres seemed like it would be ideal for our ponds.

We designed twelve nice, rectangular ponds each with a drain in the corner. The drain lines would connect to a harvest structure, which would house a shrimp pump. Glory be! We would finally get to advance to some modern technology and use a bona fide shrimp pump to harvest the shrimp. To finish off the new design, we proposed a big ditch that would carry all the drained pond water, especially at harvest time, to another part of the property, far away from the main part of the farm. This area is mostly flat but, on this property, there was just slightly enough elevation fall on the 1,400 acres to flow the pond water away from the ponds and out of the way during harvest. This part took some careful civil engineering, but we pulled it off thanks to the luxury of having so many acres of land with which to work.

After we designed the farm, the next thing we decided to do was drill a new, big water well. This was at Dan's insistence. We were unsure of the performance of the older wells and couldn't risk the possibility of not having enough water to keep fifty acres of shrimp ponds full and going. He was insistent about assessing all the risk possibilities—he saw danger and destruction in every task and every situation. For every good possible outcome, he could envision twenty ways of ruin. I believe time spent on risk assessment is a crucial aspect to the project's planning and is a step probably overlooked by most enthusiastic new entrepreneurs. Dan took risk assessment to a whole different level. In the end, his aversion to risk made us build a reliable farm and operate it better than most companies do; I just wished that the worry and paranoia didn't weigh so heavily on him.

We hired an old-timer who had a cable tool rig to drill the new well. His name was Cecil Jeter, and he was a hoot. He told stories and talked more than anyone I ever was around. He just never stopped talking and spinning yarns. He had been a cable tooler for years and years, drilling oil wells and water wells. He was also just about the last cable tool driller in the country and maybe the best there ever was. A cable tool rig is quite a thing to see and watch operate. Instead of drilling a hole in a circle with a bit like a modern rotary rig does, it just beats a hole in the ground. It does sound crazy, and it is. Just as crazy as the men who operate these things, but once you are around one you can see why. The cable, suspended from a derrick, has a huge iron bit attached to the end that must weigh 4,000 pounds. It goes up and down thunderously beating a giant hole in the ground. Each time the bit falls and hits bottom, everything shakes, even your brains. Imagine doing that every day for fifty years. You'd be crazy too with your brains bouncing around in your head. Jeter was crazy, but in a good way. He was very smart and a fun guy to be around and learn from—and I learned a lot because he never stopped talking. The best part is that water wells drilled with a cable tool rig are usually producing wells. This type of drilling sort of acts like drilling and fracking all at the same time. Each time the bit hits the formation it breaks it up and cracks it. A tool called a bailer goes down and grabs the cuttings then brings them up and out. When done properly, and if the formation is porous to start with, phenomenal water wells result with this method of drilling. Jeter drilled us a very good well in about two weeks. Dan's worries about water eased for a while so he turned his concerns to the next problem—dry dirt.

It turned out to be fortuitous that we got all the water wells pumping because the Mennonite earthmovers came to me and said that the ground was so dry that the fine clay soil was just powder and would not form into levees. It would not form into anything. When the scrapers cut and then dumped the dirt it just spread out like talcum powder. These guys had never seen anything like it before, though their experience with dirt in the desert was limited. After a few days of fighting the powdery mess and trying to force our will on the dirt to form into levees, we decided that the only way to make it stick together and form a pond bank was to use the water wells to flood the whole area. We would pump water out on the ground and let it soak until the dirt had enough moisture in it to cooperate. This additional step set us back by about two weeks, but we got that whole area of 150 acres well flooded, ready for building ponds—and build them we did. We got the ponds finished and the plumbing and electrical in and the whole shebang ready by late May. We got our post-larval shrimp without any trouble in late May. At that time, many shrimp farms had begun operations on the coast in the South Texas area as well as South Carolina and other coastal states. Since these farms were all near the coastline where the weather is much warmer earlier in the year than in West Texas, these farms had all stocked their baby shrimp in April and early May. By late May when we were ready and the weather was finally warm in West Texas, the hatchery had plenty of baby shrimp available for us to stock in our new ponds. It had been a very long marathon of solid work for those four months, but somehow we did it and Regal Farms was off to a great start.

As the owner of Regal Farms, Dan was painstaking about details and helped me add rigor and discipline to the whole operation. It was a good lesson for me. Dan came from the tech industry and brought many good practices with him that I could implement on the management side. He taught me a lot on the business side too: balance sheets, the profit and loss statement, sources and uses for spreadsheets, and a myriad of other business tools that I wouldn't have encountered otherwise. Marine biologists don't get much business training in school and Dan was a good teacher.

Dan was also weird. Really weird, in fact. When I think back on all the information and knowledge I got from Dan, I cannot get out of my head the environment in which I usually got the information. Dan lived in another state, but when he was at the farm, he and his wife stayed in a mobile home on the property, just across the gravel drive from the house I occupied with my wife and kids. Usually at the end of each day that he was at the farm, the same strange scenario played out. He would invite me over to his trailer for either a work-related meeting or just for a beer. I always accepted. After I mustered the nerve, I would go over and knock on the door to their mobile home. Dan always answered the door wearing only white briefs and a pair of black socks. Nothing else. He would sort of point back at himself and say, "This doesn't make you uncomfortable, does it?" It was uncomfortable as hell, but I would just blow it off and pretend it was no big deal. "It's your house, bro, do what you want," I would say. He was my boss, after all, and meetings were necessary and I was for sure not going to decline, no matter how uncomfortable I felt. We would then have a beer and discuss the day or have our meeting on some business subject or farm issue. Afterwards, I would go back across to my house. Every day we did this. It was notable to say the least. Later I learned that he had met with some of the shrimp buyers and at least one contractor like that. Eccentricities aside, the information and skills I got from Dan about business would serve me for the rest of my career, especially once I started my own farm. What he got from me were some very nice crops of shrimp and a profitable business. That first year in 1995, even rushing to get the whole farm built, we gathered between six and eight thousand pounds of shrimp per acre. Everyone was ecstatic—but learning how to harvest that farm was a series of hard lessons.

The first hard lesson stemmed from Dan's continuing paranoia. That year it was exceptionally cold at harvest time. Before we started draining the first pond, a process that took about twenty-four hours to complete, I turned off the aerators to the pond. That's what I always did at harvest time when the weather was cold. That year it was very cold, and the water had plenty of oxygen to keep the shrimp alive at that temperature. The aerators at Regal Farms were different from those used in other farms. Instead of a motor turning a couple of paddlewheels splashing the surface, these new aerators were basically 3 hp motors with a submerged propeller on one end spinning at 3,500 rpm. I wanted them off so they would not damage the shrimp. During harvest, as the water drained down the shrimp would become more and more crowded, and I feared that those aerators would just shred them once the water was low. Dan had a fit, screaming and stomping around and nearly having a stroke. He insisted that the shrimp would die from lack of oxygen and ordered me to turn the aerators back on and to leave them on until the water and shrimp were out of the pond. I strongly counseled against that action, but in the end told him it was his farm, and he was the boss. Well, I was right—it was a very bad idea. When the pond was finished and the last of the shrimp and water had left through the drainpipes there were mounds of chopped up shrimp near every one of those aerators. I bet each aerator destroyed 500 pounds of shrimp, and there were five aerators in each pond. It was a stupid and expensive mistake, but that's what unwarranted fear and unreasonable worry can sometimes produce.

Next on the learning curve for harvesting at Regal Farms was dealing with stranded shrimp. The way we built these ponds was with the drainpipe on one end that, at harvest, attached to a shrimp pump which carried the shrimp up and out of the drain ditch and into the waiting baskets of workers located on the opposite levee. On the other end of the pond was the water inlet pipe where water came into the pond to add water. We noticed from harvests on the other farms that the shrimp would follow the slope of the pond down to the drain as the water level lowered, but when clean water was introduced from the inlet pipe (as needed during harvest when the water got low), the shrimp would turn around and walk or swim back up the pond into the clean and warmer water. This caused a mess because just when you thought all the shrimp were concentrated at the drain end and would soon be out of the pond, those jumbo shrimp would turn around and go back. Now you had shrimp all over the pond bottom once again, but with so little water left, they would get stranded and have to be picked. Water from the inlet pipe had to be used because the pond water became so muddy and nasty in the deep end that if you didn't run some good water down to the shrimp, they would perish in that mess before you could get them through the drainpipe and into the pump. On top of being clean, the well water was a balmy 70° and with the weather so cold at harvest time the shrimp wanted to be in that warm, clean water.

To fix this issue, we eventually purchased large rolls of lay flat poly pipe that was 16-inch diameter. This material is used by farmers to irrigate crops on the cheap because it's just made of thick plastic, like heavy garbage bags, formed into a long tube. We clamped this tube around the inlet pipe and waded through the pond rolling this pipe out until it ended at the deep end, just in front of the drainpipe opening. This way the clean and warm water travelled through the poly pipe and dispersed right amongst the shrimp, keeping them happy and in the location we wanted them. It worked like a charm, and while we had stranding troubles again from time to time, we never had them because of this issue. Before we developed that solution, however, we had to deal with stranded shrimp all over the bottom of the pond.

On that first pond, we stranded so many shrimp because of the water situation that we had to go round up whomever we could find in Imperial and Grand Falls to pick shrimp for us. We could usually find five or six willing souls to come work. Of course, many people had regular jobs, making it hard to find good hands on short notice. Most of those we did find were eager to work. It was an experience to work alongside these folks. At the end of that first day, everyone was covered in mud and sweat, and we were all so tired we could hardly stand. One of the pickers said, "Damn, this is the suckiest job I ever had. Way too much dirty work. Why would anyone do this?" I told him, while at the same time reevaluating all of my life choices, that I had gone to college and gotten a master's degree so I could do this crappy job. He left that day and never returned. But most stayed. We learned who of the lot could do the jobs requiring a little more responsibility and trust and we learned which folks just needed to be in the mud picking shrimp. The faces would sometimes change but their circumstances were usually similar. As long as they could show up somewhat sober in the morning, ready and able to pick up shrimp and carry those full baskets out of the pond, that was all we cared about.

We also learned that as the ponds drained at night we could use very bright spotlights to herd the shrimp from the shallow end down to the deep end. Shrimp are very sensitive to light and will immediately jump and breach the surface when hit with bright light. Then they swim away from it. We would use this to our advantage to hustle them down to the deep end and that would help reduce the stranded shrimp in the shallows. Shrimps' sensitivity to light can also be a problem in West Texas. Often during the growing season there would be harsh nighttime thunderstorms, after which we would find shrimp all over the levees where they had jumped from the flash of the lighting, breached the surface, and then were blown out of the water by the blasting wind onto the dry land. Those shrimp were picked also but usually for our personal consumption.

I would be remiss if I didn't talk about three other folks who were paramount to the harvesting success at Regal Farms and later the Permian Sea Shrimp Company and others. These three were A. C. and Judy Stephenson and my wife Patsy. A. C. and Judy were an interesting retired couple who had moved to Imperial about the same time Triton Aquaculture started. A. C. told us that he had a career doing some sort of classified work for the US Navy during the Cold War. He never gave us much detail about what he did before retiring to Imperial. We never pressed him. They lived a very interesting life together all over the world and were always thirsty for new things and adventures. Ever since they moved to the Imperial area they had been fascinated by shrimp farming. A. C. and Judy had their own alfalfa hay farm to tend to, but they continually asked if they could somehow participate and help in some way. I appreciated their enthusiasm but felt it would not be appropriate to ask a couple well into their 60s to help pick shrimp or do any of the hard work necessary at harvest time.

They practically begged us to let them come pick up stranded shrimp on harvest days, so I relented and invited them to come help. They would end up coming to every harvest on every pond for the next ten years. We would give them our fall harvest schedule and each morning they would show up with coffee and some home-made delicacies for all to eat and enjoy. Often, they would bring me a bottle of something strong to stash away for an after-harvest celebration. Once they finished distributing their goodies for everyone, they would hit the bottom of the pond, baskets in hand, and help hunt down every delinquent shrimp in the pond. They never missed a harvest day and they never once asked to be paid or compensated. For them this was a fun and an integral part of the unique desert shrimp farming culture that developed. This was how they lived their lives to the fullest. It did, however, always bother me that they worked so hard for me for free. I finally did get to even the score by helping them operate their hay farm in later years when they got too frail to drive tractors and haul hay. It was the least I could do to repay them for picking up hundreds of pounds of shrimp throughout the years.

My gal Patsy was the other critical part of all this. She would help call and coordinate the pickers on harvest days and then she herself would bail off into the ponds to do her share of the picking. She would handle the local retail sales and make sure people got their shrimp and the farm got its money. She also made sure everyone got fed during harvest days and that there was plenty of water to drink to avoid risk of any heat stroke or dehydration. She might very well have been the most important hand we ever had. She never got paid a dime, either. Patsy stayed so busy on harvest days that in order to keep our baby Hannah occupied she grabbed a shrimp basket and plopped our little one inside, creating a makeshift playpen. Then she'd throw a few live shrimp from the shrimp pump into the basket for Hannah to play with while she got on with the work of the morning. I bet Hannah is the only kid in America who teethed on the tails of live jumbo shrimp!

On top of all the work to get those stranded shrimp out of that first pond, the buyer at the time tried to take advantage of Dan's inexperience and pull a fast one on us. Inexperienced as he was with shrimp farming, Dan was appalled that shrimp would strand in the mud, and before all that mud and junk was washed off them, they looked terrible. A good hosing down easily remedied the situation, but Dan was convinced that all the stranded shrimp were wasted.

He could not imagine that anyone would buy the shrimp that had been covered in mud and was ready to write off all the stranded shrimp as lost for good. We had 14,000 pounds of clean shrimp in the truck and Dan was ready to take a check for them and be done. The buyer told Dan that he would do us a favor and if we would just pick the stranded shrimp, wash them, and load them up he would haul them off for us. Dan was ecstatic that he would do us such a favor. I growled in protest to Dan and the buyer. I told them both that the buyer knew full well those stranded shrimp were part and parcel of shrimp farming (and muddy shrimp sometimes are a part of wild shrimp catching too). I told them that after a good washing the buyer would get just as much money for them as he would for the others that harvested clean. I could see him almost smiling at Dan's naiveté and it was making me upset. Of course, Dan as usual wouldn't listen to me, but I think the buyer knew he might not leave there without a word from me if he kept the charade going, so he offered to purchase the picked shrimp at half price. Dan took the deal and was just positive he had swindled the buyer who, I'm sure, broke a rib laughing all the way back to Brownsville. By the end of that first year Dan finally learned that, unless there were other factors with the picked shrimp like heat exposure or physical damage, they would bring the same money that the rest of the shrimp brought if they were properly cleaned and iced. Some people just have to learn the hard way, and Dan was definitely one of those people.

The year I left Triton Aquaculture, they had also made a crop of their Lost Sea brand shrimp, sold them to a buyer on the coast, and then closed their doors for good. Now Regal was the only farm in the area. The industry needed more investors and more trained shrimp farm managers if it was going to develop. Being exclusive often is an advantage but not necessarily when you're a shrimp farmer all alone in the desert. The coastal buyers would be more competitive if there were more options and more ponds to bid on and purchase, which would help their bottom line and the farmers' as well. If enough farms operated in the area, then maybe one day we could justify a processing plant and not have to sell to the coastal buyers—control the shrimp from farm to plate. For me personally, it seemed cool that a whole industry might develop out of all the hard work and ideas. Those were worthy goals, but mostly I was looking forward to having other shrimp farmers and the like to work and socialize with. It was hard and lonely out in the desert, doing something no one in the area could relate to or understand. Most of the locals thought the whole shrimp farming idea, and anyone involved in it, was a little nuts.

Crazy idea or not, it was working. In 1996, Regal Farms produced more shrimp than any other farm in Texas, even coastal farms five times its size. It was an amazing and impressive feat but not because of why you might imagine. We did not harvest 20,000 pounds per acre or raise the whole crop to some jumbo size that sold for a fortune. We met our average of 6,000 to 8,000 pounds per acre, which was great. We consistently pulled those sorts of numbers. Sometimes a pond or two might get close to the magical 10,000 pounds an acre, but we stayed consistent and made nice crops of large 21–25 per pound and 26–30 per pound tails, which is the shrimp count in a one-pound bag. How then did we gain such an honor as the highest production in Texas that particular year? Simply because no other farms managed to make a crop, or not much of one, except us that year. There was (and still is, lurking somewhere out there) a deadly virus for shrimp called the Taura virus that made its rounds throughout the Western Hemisphere in the early nineties. It turned out that 1996 was the year it decided to summer in Texas. The virus literally wiped out all the farms on the coast. Piles of dead shrimp washed up each morning in every one of their ponds. It went on and on, all season long. All the coastal farms looked and smelled like death, and all the owners and workers felt the same. The Texas coastal farms produced less than 800 pounds per acre that year. Most of the farms produced nothing, not one single live shrimp. West Texas was another story. We had a banner year and produced as good as or better than the year before. The general consensus at the time was that the virus was carried by the larval shrimp from the hatcheries that must have been infected with it. The thought was that each farm became infected the second they put the baby shrimp in their water. The strange thing was, we got our larval shrimp from the exact same hatcheries the coastal farms did—even on the same day from the same larval rearing tanks in some cases—but we never saw a hint of virus-related mortality. Our shrimp were fine and thrived as always.

It is my feeling that the water in the desert is protective against the virus and some other pathogens that infect shrimp. Our water is the same salinity as the coastal bays, but the mixture of salts is different. The one extremely high component in the Pecos Alluvial Aquifer is gypsum, or calcium sulfate. I believe the gypsum and other calcium salts somehow protected the shrimp from infection. It could also have been a whole mix of chemistry and not just these two items, but either way, there was something going on with the water that protected against the virus because our farm should have been affected if it was in the baby shrimp. Water chemistry has been found to affect the infection rates of certain bacteria and viruses on algae species, so there is precedent for this sort of protection in nature. Once again, it is another amazing attribute of this local resource that makes West Texas such a great place to grow shrimp. It was a thrill to be the highest producer in a state that had a track record for producing a lot of farmed shrimp, even if it was kind of a crappy way to win the title—like winning a game because the other team had to forfeit. I celebrated anyway, though quietly, just out of courtesy to my shrimp farmer friends on the coast who nearly lost everything.

MY BRUSH WITH DEATH

My silent celebration of that state record shrimp crop was especially sweet since just before harvest season I had a scary semi-brush with death that I was lucky and blessed to have lived through. I am going to tell this story since it was caused by a farming-related illness and one that often affects folks who work in agriculture. It is a job hazard that is especially risky for farm and ranch workers, and it is almost always fatal. One day in August started as always with breakfast and conversation with my wife and kids before I headed down the gravel road to the ponds. It felt like any other day, and I was ready for the typical managing of the farm and seeing that things were working as they were supposed to. It was just about five or six weeks before the start of harvest, so this was also a critical time for the farm. By mid-morning I was beginning to feel like I had a bit of a cold coming on: runny nose, headache, and maybe a bit of a fever. Weird, as it was August and hardly cold and flu season. By lunchtime, I was sick. I felt like I had the full-blown flu. By afternoon I was shaking like it was twenty degrees below zero, sweating like I had run a marathon, and hurting all over. I hurt so bad, and I could hardly hear anything. I also had little rash bumps covering my body and my lymph system was swelling under my armpits and in my neck. In addition, I was vomiting every five or so minutes. It was so miserable I was praying for death.

My wife insisted we go see a doctor, so she helped me into the truck and took me up the road about thirty miles to Crane, Texas, where there is a rural medical center next to the hospital. This place was always where I went—albeit rarely—to get medical attention. A few nurses staffed the center, and the doctors there were part of a circuit of doctors that traveled around and saw patients throughout rural West Texas and the Southwest. I had hoped that I would just die on the thirty-minute trip there, but since I didn't, I hobbled out of the truck and, after puking in the parking lot, I managed to get inside the door to the waiting room. I was lucky there were only one or two others in the waiting room at the time so in short order they helped me to an examination room. I think I passed out a couple of times while waiting. After a nurse took my temperature and asked all the regular questions, an older and very weatherworn doctor came in to assess my condition. This guy seemed to be very battle hardened and had that wise but mean look about him. He asked me what was wrong, and I told him I was sick. He laughed and told me I didn't know what sick was. He said, "I saw people in Africa bleeding out of their eyes from Ebola and people in South America with flesh-eating bacteria that made their body parts fall off. You ain't sick, you just feel bad." Well, damn, I thought and said, "Okay, sick or not, I feel REALLY bad."

I think he had a suspicion, but he went ahead and asked me the pertinent questions in order to be sure: Did I work on a farm or ranch? Did I work in or around barns that hold feed for animals? Were there ever any signs of rats or mice in those barns? Of course, the answer was yes to all three questions. There are always rats in barns and especially if there is animal feed stored there. He was silent for longer than I liked so I asked him, "Do you have any idea what is wrong with me?" He answered—no kidding—"I think I have an idea, but we will know for sure when we do your autopsy." Where did this guy learn his bedside manner? Maybe he had been a prison doctor or a doctor for a drug cartel. I think my wife passed out at that point. He followed up by saying, "It looks like you have a Hantavirus, you know, the rat virus, which is common in the desert southwest, and it's got a low survival rate, so in the next five days you'll either get better . . . or you will die. I will prescribe some things to make you comfortable and bring down the fever but there is nothing we can do to cure this virus. Call me in a few days if you are still alive." That is exactly what he said to me, word for word. It turns out I lived after all, but only after about three more days of suffering. I never called that doctor back. I hope he still wonders if I survived. Agriculture is one of the most dangerous professions in America. That is why the federal government lists statistics for work-related accidents and deaths as non-agriculture industries or occupations. If they included agriculture, this list would be very long since there are many deaths and injuries from farm animals, farm equipment, and weather, not to mention deaths from exposure to rat excrement.

While we are on the subject of death and dying in rural West Texas, I will describe what I believe was the most bizarre and comical funeral I ever have or will ever again be so privileged to witness. Regal Farms had on its west side some very interesting neighbors consisting of a woman named Rhonda, her husband, and two sons about junior high age. They had purchased a couple of sections of land to raise livestock and live off the land. Being of very modest means, it apparently took all their savings just to purchase the land and thus they had no permanent dwelling in which to live, so they lived in a tent camp near their animal pens, close to the fence from the shrimp ponds. I never actually met the husband or even saw him around. I understood that he was an electrician, and he commuted each week to work somewhere back in East Texas, the area they had moved from, leaving Rhonda to tend the animals and the ranch. She often came to the shrimp farm to give us some help or assist in harvesting. She was quite a gal—rather large, well weathered, strong as a bull and always at work with a revolver in a belt around her waist. And this was before wearing a sidearm was in style like it is now. She was always dirty from work and smelled of cigarette smoke and pig droppings but considering they didn't have any visible means of bathing or showering, it could have been a lot worse. She was a pleasant person to be around and seemed to be terribly happy living in rural West Texas as opposed to wherever they had come from. She worked hard every day tending to her animals, mostly pigs, and taking care of the tents and pens. She was by all accounts a good neighbor, and we tried to be good neighbors as well. Whenever she requested our assistance, I never refused and always made sure our hands were available to her if she needed them. I was in awe that they could live like that and be happy all the time. It made me really appreciate all the blessings that I had.

I guess the electrician business was prospering back in East Texas because, lo and behold, a nice, new double-wide trailer house with air conditioning, running water and all showed up one day. In forty-eight hours, the tent camp was disassembled and all the household items from around the pens were collected and the whole show was moved into the new house! After two years of living in a tent next to a bunch of pigs, it must have been nirvana. The house was quite a bit farther to the south on their property, so I saw very little of Rhonda anymore except at the post office or gas station from time to time. Just a few months after the double-wide was moved in, we got word from Mr. Dean, a preacher by calling and oil field pumper by trade, that Rhonda's husband had died. My wife and I had run into Mr. Dean at the post office, and he informed us that Rhonda's husband had just driven in from working all week and dropped dead the next morning. He asked us to pray for the family and whether it was possible to help with the funeral as they had no money and were dead broke. I guess the new house expenditure had drained the last of what little they had. Mr. Dean is a fine man and truly has a heart for his neighbors. To this day my son is convinced he is an angel like the ones the Bible says we entertain without knowing it. He is one of the best men I have known, and he wanted to help Rhonda through this tough situation. We said we would do whatever we could.

It happens that there are a lot of interesting facts on many subjects that are known to those in rural America that are not so commonplace in the cities. One of these secret folk wisdoms is how to make paupers' funerals much more cost-effective in light of the relaxed laws and cultural norms out in the sticks. This can come in handy since a lot of rural people have very little means or use for a fancy send-off. Part one of the secret bargain funeral knowledge is that you don't, by law (at least in Texas at the time), have to embalm a body unless the person had died from a communicable disease. This makes the whole thing much more economical because embalming is about 50 percent of the whole cost. Without that, the only thing really needed from an undertaker is the use of his cooler to store the body for a few days and if you move fast, you don't even need that. That fact is important to this story. The second most useful thing is a cardboard prison casket. These are available from the local prison for next to nothing and come as a little cassette that you pop up and put together much like a shipping box, but UPS won't come pick it up for sure. They are just strong enough to hold a body, but they are simply made of treated cardboard usually painted white. Since the coffin is the other major expense of a proper burial, this will save a fortune. Some folks obtain them in advance. A properly pre-planned cheap rural funeral will save vast amounts of money. And lastly, there needed to be a place to put the dead person in the ground. This is often a plot in a local cemetery, which in rural areas is usually cheap; if not, there is a handy remedy. You can obtain a permit from the county to bury a body on your own land, which is exactly what Rhonda did. She got a permit to bury her husband right smack in the front yard of her brand-new double-wide trailer house, not fifteen feet from the front porch. I am sure some folks might find that awkward, but love is love and it was his hard work that was responsible for the new trailer in the first place; it is only fitting that she wanted him to be close to home, forever.

Since a few of us neighbors were willing to help, we started getting things ready. It was a Saturday morning when the death occurred, and we were trying to move fast to get this deal done since the body would not have the benefit of embalming and time was of the essence. Ronnie from the water district brought his backhoe over and dug the grave in the exact spot Rhonda pointed out to us in the front yard. Once a proper hole was dug, we set three two-by-fours over the hole, which would support the casket during the funeral. We also laid two long heavy ropes over the hole to lower the casket after the funeral when the boards would be slid away, and the casket could be lowered down by a few of us. We set a large tarp over the area, laid out nice and flat to cover the hole. An open hole is very depressing, and it is typically not something the family wants to see during all the eulogistic words and memories. I thought it looked nice and respectful. We felt that even though it was simple, the event could still be reverent. Now we figured that this funeral was going to be happening the next morning but were informed that the extended family was coming in from, I assume, back in East Texas where Rhonda and her husband had come from when they moved here, and we were to wait for them. Imperial is a long way from anywhere and it took them a day or two. The funeral was moved to Tuesday morning at daybreak.

We all had decided to meet Mr. Dean at 8:30 a.m. for the funeral. I dressed in my nicest jeans and a starched shirt, my best Sunday boots and hat. My wife also dressed nice and respectful as did all my hands from the farm and the handful of folks from town. We all got to the site and assembled sort of off to one side so as not to crowd the family at their time of mourning. They had not arrived yet and it was pushing 9:30 a.m. and getting a bit warm. The prison coffin had arrived about the same time we did so Ronnie, Dean, and I placed it on the boards and ropes, over the grave. We tidied up the tarp so the whole thing looked like a proper funeral gravesite. It was now 10:30 a.m., and still no family. Rhonda finally came out and indicated that they would be out in just a bit. We figured they must really be getting dressed up for this thing. Just before 11:30 a.m. we finally saw movement from the front door of the house. Out came all the family pouring onto the front porch—dressed in cutoff shorts, flip-flops, and wife beater tank tops, the men and the women. They had apparently been drinking all night and were still at it, as every one of them was blitzed out of their minds and laughing and hooting like this was a ride at Six Flags. They all stayed on the house porch, as I doubt they could negotiate those three steps leading down into the yard in the state they were in. Since the whole thing was happening right there in the front yard, they would be able to hear Mr. Dean deliver his heartfelt eulogy.

Dean started his little sermon and send-off. He is a very good preacher and took this just as seriously as he would a funeral at the town cemetery. The revelers on the porch paid little attention to the words Dean said. They were busy drinking, laughing, and partying. The rest of us were trying to be respectful and not indicate how appalled we were, so we listened and focused on Dean's oratory. It also seemed that the family had brought with them a large dog, a German Shepherd—I had never seen this dog before, so I assume he came with the extended family. In the midst of Dean's sermon, the German Shepherd starts wandering around in and out of the crowd. All at once, he freezes like a statue, nose in the air and a look of resolve across his face. Seems he was now downwind of the casket, said casket containing a four-day-old body lying in the 100° West Texas heat. The dog went straight for the casket, and once he hit the tarp it collapsed and into the hole he crashed.

Mr. Dean was trying to have a serious funeral here and never missed a line. The family revelers found this terribly amusing and really got to hootin' and carrying on. Meanwhile one other fellow from town and I tried as discreetly as possible to get the dog out of the grave and end the whole kerfuffle, but he just kept going right back in trying to get to the old—and ripe—ex-electrician. Finally,

Mr. Dean realized that no matter the effort, this thing is a cluster of monumental magnitude. The dog was in the grave trying to paw the casket open, and the family was seemingly uninterested in being cooperative or having a serious funeral, so Dean just brought it to a close. The family went back in the house to revel and drink. By this time, I had tied a rope to the dog's collar to control him, so I handed the rope to one of the family and asked them to take the dog inside so we could finish up. Ronnie and I and a few others wasted no time getting that tarp out of the way, pulling out the boards, and lowering that casket into the grave. Fortunately for us, the smell that caught the dog's attention was still too faint for us to pick up. Ronnie got on the backhoe and got the grave covered up in record time. I was appalled at the whole thing. Later that night, after a few beers we all just laughed and laughed. All I know is I never want to be a part of a funeral like that ever again.

A LESSON IN (UN)ETHICAL BEHAVIOR

That second year was also significant because it was the year Dan and I changed shrimp buyers. This new fellow would eventually become a good friend, even to this day. Ultimately, this adjustment was a good thing, but it was wrought by many unpleasant incidents via another episode in poor judgment and decision-making on Dan's part. Of all the positive business lessons I learned from Dan, on this occasion I learned a lesson about the worst intentions and behaviors of men. This experience too would serve me the rest of my career. I learned that many people let some things worry them or concern them the most because they themselves are willing to perpetrate those things; therefore, they are paranoid that everyone else will perpetrate them also. Sort of like being certain that everyone else cheats on tests because you do. Dan was extremely paranoid and worried that our first shrimp buyer, who for the most part was fair and honest (stranded shrimp aside), was cheating us somehow.

He was the first coastal shrimp buyer that took a chance on West Texas shrimp and gave us a shot. He also had great connections in the seafood industry and always gave us good prices. He tried to play the game, as all buyers do, saying, "Oh, in this pond you're about to harvest, the shrimp are a little darker than that last pond, so I have to come down twenty cents" and comments like that. They all do that, and you just have to stand your ground and play the game in your favor.

Overall, this guy was a straight-up dude, but Dan was certain he would rip us off sooner or later and break the contract we had negotiated with him at the beginning of the harvest year. Dan just knew it. I could never figure out why because the buyer never gave us any indication that he would do that to us. I think Dan was paranoid about it because he himself wanted to break the contract—or at least knew he wanted to try. I believe he, in his inexperience, thought this was how businessmen get rich. I know a lot of naïve people think that way, but it is generally not true. Even in a world as messed up as our world is now, people that do not honor their word and their contracts are pariahs to most other businessmen. Here is how the whole sordid mess went down.

It was very late in the fall and the second-year harvest season was well under way. We had harvested about eight of the twelve ponds and were preparing to harvest pond number nine. By this time, the weather was getting colder with the possibility of shrimp-killing cold temperatures arriving at any time. It was also deep into the shrimp-harvesting season on the coast and the market was getting oversupplied and soft. Our buyer let us know that he was going to have to start purchasing our shrimp at the low end of the negotiated price range due to market conditions. Dan was upset, and he was just sure that the buyer was paying the coastal farms far more than he was paying us. Actually, he probably was since he didn't have to truck coastal shrimp eight hundred miles, but that issue was part of the negotiations that led to our contract agreement and all that trucking expense had been factored into the pricing. Dan knew that, or should have known that, but no matter, he was walking the floor worrying that we were getting ripped off and at any moment the buyer would just quit the contract and abandon us. So, he hatched a plan.

Now I had no idea of this plan, or my actions would have been much different and would at the least have led to much less hard work, but I'll get back to that in a bit. As I mentioned before, each harvest took about twenty-four hours to complete. We started draining water off pond number nine at about 6 a.m. one nice fall morning. The truck that had ice and totes was supposed to leave from Brownsville about 10 a.m. and head to the farm. It would take ten hours or so, give or take, for the truck to arrive. By the time the truck made it to the farm, the pond would be about half drained and we would be ready to start the shrimp pump and collect shrimp. The truck had to be there when we started running the shrimp pump so that we would have ice and totes in time for the shrimp. No truck meant no starting the shrimp pump. So, we started draining that pond right on time at 6 a.m. Dan never said anything that would make me think anything would be different. He waited until he was certain the truck had left for West Texas, then he called the buyer. He insisted that the buyer pay a better price, or he couldn't have this harvest or any other harvest the rest of that year. We had no backup buyer at that time, so it was a very stupid bluff. It turned out the buyer would not budge, and he was upset that Dan might try a stunt like this. I never knew exactly what happened, but he called the truck back and either told them to turn around or not even leave, but unbeknown to us, there was no truck coming. In Dan's arrogance he was sure that he had the buyer over a beam. He just knew that truck was coming, and the buyer would have to cave to Dan's demands and pay a better price, not only for this pond but for the rest of the ponds as well. Dan came to the ponds and told me of his so-called brilliant plan. I was dumbfounded by Dan's unethical behavior and lost all respect for him that day.

I always kept an eye on the weather, especially on harvest days. The weather can change everything, and this time was no exception. By that afternoon, I heard on the news that a massive cold front was headed our way and would hit about four in the morning. My calculations showed that the pond would be just about finishing at about that time, and I figured the cold air would actually help cool down the last of the water and the shrimp would run from that cold water and move right out the drain. At least that is how it should have worked. When the truck had not arrived by 8 p.m. I called Dan to ask when the truck would be here. He didn't know but assured me it would be in soon. I told him we needed to know for sure, but Dan and the buyer were still arguing about the contract and the buyer never told him that he had held the truck back, so we had no idea. Finally, after more waiting, I told Dan that the pond was more than half drained and a hard front was coming; with no truck at this critical time, it would be a good idea to abort and start filling the pond back up. It would take hours and hours but at least the well water was warm, and we could start the harvest again the next day. Dan refused and insisted that he had control of the buyer, and we were on for this harvest. He insisted we not change anything, so we kept draining.

Just before midnight, Dan came to us with the news that he had finally made a deal with the buyer, but he had also learned that the truck was just then leaving Brownsville and it wouldn't be at the farm until about nine or ten the next morning! I was livid. We could not refill the pond even if we wanted to at that point, and since the truck was so late coming, we would need to finish harvesting the pond later in the morning when the truck arrived. This meant that all the shrimp would have to just sit in that pond, already two thirds empty of water, while that hard cold front hit. It sounded like a disaster in the making, and of course it was. We just sat there and waited, and the front hit and the pond water temperatures fell so fast that all the shrimp just stopped in their tracks, lay over and died. There was not one live shrimp left in that pond and we had not even started the shrimp pump yet. They were everywhere on the almost 4-acre pond bottom—all of them lying dead under about one foot of ice-cold water.

Dan went back to the house to rest from his triumphant negotiations while we waited for the truck to arrive. Once we saw the truck pull in, we began to drain the pond again and started the pump. We got virtually nothing through the pipe and into the pump. As the water receded into the drain ditch and the bottom of the pond slowly started showing, we could see why. We could see the nightmare. Dead or cold stunned shrimp everywhere! All we could do now was call the pickers and tell them to bring their friends and families. It was horrible, but at least the cold would preserve all the shrimp for many hours while they lay there waiting for us to collect them. We picked up 12,000 pounds of shrimp, one 20-pound basket at a time. We picked for hours and hours and hours. Every basket of shrimp had to be power washed to get the pond mud off the shrimp. Finally, after ten hours, the truck driver said he had to leave with whatever load we had, period. We estimated we had to leave about 7,000 pounds of shrimp in the bottom of that pond to rot. I told all the pickers to take home all the shrimp they could carry from the leftovers. After the truck left, we let all the town folks know they could come get whatever they wanted out of the bottom of that pond before it spoiled. At least the cool conditions kept the shrimp in good shape while people came and got them for the next two days. Many folks came from Imperial and the nearby town of Grandfalls to glean the shrimp. Everyone wanted to stock up on shrimp, except our buyer from the coast. At Dan's insistence, the buyer agreed to take that load, but after that truck left, the buyer tore up the contract, told us to go screw ourselves, and never, ever bought another shrimp from West Texas. Thanks, Dan, lesson well learned.

We still had four more ponds to finish for the 1996 season, but we had no buyer, and our owner now had a terrible reputation in the industry. Fortunately, lots of folks in the seafood industry—from producers to buyers and processors—have terrible reputations. It's so common, the folks who don't have bad reputations stand out from the crowd as an anomaly. It almost becomes a contest. If you have a bad reputation, it's no big deal because someone else has a worse reputation and is not intimidated by you at all. May the best scoundrel win. It wasn't long before we had a replacement buyer in place: Dennis. We were pleased to have Dennis come and purchase the rest of our shrimp, and the prices he offered were fair. What was not good was the way he wanted shrimp iced and toted. Our previous buyer had a great system for icing and toting up shrimp. He would send a truck with large totes that were about half full of ice made with salt water. Ice made like this was super cold and slushy like a snow cone. As the pond drained, the pump would direct the shrimp into a dewatering box. The water drained off the shrimp, and the hands would open a gate and shrimp would pour out, filling shrimp baskets with about twenty-five pounds of shrimp. We would weigh the basket and dump the shrimp into the slushy tote. Basket after basket would fill up and the procedure repeated itself. Every now and then we gave the shrimp and slush ice a good stir, and when the big tote weighed about 1,500 pounds, we would pop the lid on it, load it in the truck with a forklift, and start filling the next tote—easy-peasy. We were spoiled.

The system Dennis had was much more labor intensive and made harvesting shrimp a miserable experience for everyone. His system was like how the old shrimp boats used to ice and store their shrimp—ice shoveling and manual labor. His truck showed up with a wall of ice filling the truck and a large supply of small plastic totes that held 100 pounds of shrimp each. You take an ice shovel, shovel out a scoop from the wall of ice and put it in the little tote. You dump in your basket of shrimp, another layer of ice, another layer of shrimp, and so on until the little tote is full and the last layer of ice is covering the last layer of shrimp. Next, you stack the boxes in the truck from the bottom to the top and all along the width of the truck. As you dig out the ice, it makes room to stack more boxes and you slowly work your way through the truck until all the ice is shoveled and all the boxes are neatly stacked from floor to ceiling and front to back, where the ice used to be. This process sucked on a grand scale, and it required a lot more people to work it, which sucked even more. But thanks to Dan, for the rest of that season we had no choice. Dennis was the only buyer who would put up with Dan, so like it or not we had to do it this way. I think we were in pretty good shape after swinging those boxes up six layers high in those trucks for the remaining month. Fortunately, that was the last year we ever loaded shrimp like that, and no one was sad to see the process change, that was for sure. It would be large totes and slush ice in the future if we ever wanted anyone to work for us again.

Despite the setbacks and poor investor decisions, West Texas was now producing as good as or better than the farms on the coast, only the shrimp was different on a culinary level. Our shrimp tasted almost like lobster—very sweet and mild with none of the taste people describe as iodine flavored, the strong shrimpy flavor that most people are used to. What most don't know is that the taste they associate with shrimp is mostly a combination of rot and sulfates that are used to keep the shells from turning black. A fresh West Texas pond shrimp is much different. It still tastes like shrimp, the way you want a shrimp to taste just before you bite into a wild-caught shrimp and get let down by the experience. West Texas shrimp are fresh, sweet, and mild with a nice meaty texture. Even Dennis, our seasoned buyer, who travelled up and down the whole coastline and bid on every farm-raised shrimp pond in Texas, always said they were the best shrimp he ever tasted. He bought and sold shrimp from Mexico to Canada and every place in between. If anyone ever knew good shrimp from bad, it was Dennis. He liked them so well, he and a partner ended up buying Regal Farms from Dan after a few years of buying our shrimp.

Regal Farms produced a lot of shrimp and was the first farm to really gain the respect of the processing plants, buyers, and even other farmers on the coast. We sent all our shrimp to the coast, and they never ceased to impress even the saltiest of coastal shrimp dealers and plant operators. By 1997, we had consistent great production of more than 8,000 pounds per acre and we produced nice-looking and good-tasting shrimp. We had worked with Nutrena Feeds to develop a diet specifically for West Texas shrimp ponds and had learned how to grow good crops of shrimp with minimal water use. With consistent results and no diseases, we were the real deal now. We were even mentioned in each issue of the Shrimp News International news journal, which was a twice-yearly publication that reported on the shrimp industry worldwide and all the major players in it. We were flattered to be interviewed and mentioned each year. I was proud that West Texas was starting to be recognized as a legitimate shrimp-farming region.

We also got the attention of other areas of the country that had salty groundwater. Apparently, salty groundwater is not that uncommon. We took lots of calls from folks who had a salty well in Kansas or an old salt lake in Nevada or salty springs in Alabama who all wanted to know how they could grow shrimp at their place. They were sure they could make a go of it, just like we were. There is a lot more to successful shrimp farming, however, than having a water well that has a little salt water in it. And the work . . . no one is ever ready for the work involved in shrimp farming—working in extreme heat to extreme cold, picking 6,000 pounds of very cold, immobile shrimp from the bottom of a pond one by one, or fixing broken aerators in four feet of water in the middle of the night. You're never ready for that kind of work, and all these folks calling us interested in shrimp farming had no idea what was involved.

We did end up helping a family out of Arizona, well established in the farming business with some land and water that we felt was suitable, the Woods family. They wanted to know everything. They seemed competent enough, and since they were already involved in the agriculture business, we felt they would probably be able to pull it off or at least give it a fair go. We spent hours talking to them on the phone, and they visited the farm to learn firsthand how to do what we were doing. They also had a Texas A&M shrimp scientist come out to Arizona and help them. They got good information from the university, but they got their real-world practical knowledge from us. They ended up starting a shrimp company on their family land near Gila Bend, and branded their farm-raised shrimp Desert Sweet Shrimp. They were initially successful and even opened a couple of restaurants to sell their products. A few other locations in Arizona, Mexico, Alabama, and various other places attempted inland shrimp farming. Some were successful. Regardless of the individual outcomes of those ventures, inland, brackish water shrimp farming was now legitimately established, and thanks to Regal Farms, it was also respected and praised by the whole Texas seafood industry.

I worked for Regal Farms for three years and by this time I was really itching to start my own farm. I felt sure the only way I would ever make any real money and be able to do many of the things I wanted to do was to start my own farm and work for myself. I had started doing some interesting things to advance production while

I was at Regal Farms. I had received permission from the state to allow us to grow the Pacific blue shrimp *Penaeus stylirostris*. We tested them one year in a few ponds along with our other Pacific white shrimp and they increased our yield by about 15 percent. They also increased our count per pound by 10 percent. It seems that when you stock a different species at a low rate, like 20 percent of the population, this alternate species will grow as if it is the only shrimp in the pond. They will have high survival and grow to a larger size. I tested it with pink shrimp *Penaeus duorarum* stocked at low levels with Pacific white shrimp back in graduate school with great results and I proved my theory again here with these blue shrimp. This technique of stocking two species could really be a game changer. I also got a research permit to grow the cold-tolerant Chinese white shrimp *Penaeus chinensis*, which I hoped could get us a winter crop, giving us two crops per year. That would greatly increase the profitability of shrimp farming.

Even though the potential for tremendous production improvements with these two species was very promising, Dan was risk averse to working with these species or doing anything radical and new. He was not interested in taking any chances or doing any developmental field trials. I could see so much potential for aquaculture in West Texas and thought we could really open up our industry with these tools. No farm on the coast would ever be permitted to use these shrimp species because of the possibility of escape to the wild, which might affect native shrimp populations. The state would protect those native shrimps above all else, but because we were 800 miles from the coast there was, rightfully, little concern regarding our impact on Gulf of Mexico shrimp. We had opportunities that no one else had. We had the best water, which was much better than any water on the coast or in other inland areas, and we had the infrastructure left us by the oil fields with good roads and electrical power everywhere. We had a mature shrimp industry on the coast that would buy whatever we could produce, and we had the possibility of producing new species that could make us untouchable in the industry. I was bursting at the seams to make this venture even bigger, and I was not interested in waiting for others to get with the program. Regal Farms and I parted ways so that I could proceed with making the West Texas shrimp industry flourish in ways not yet imagined.

Just like Triton, Regal lasted a few more years and then closed down. My friend and shrimp broker, Dennis, and his partner bought the farm and ran it for another couple of years, but they could not make it work, running it from the coast where they lived. Like the lesson learned with the Imperial Shrimp Company, you just can't run these farms in absentia. In this case, trying to keep employees honest and the shrimp in the ponds and not in some back alley black market was too much of a hassle. With all their other business on the coast this headache wasn't worth it, so they shut the doors. I, on the other hand, was doing great and the future was looking good.

A FEW THOUGHTS ON THE TRANS-PECOS

I want to stray off the path here for a minute and ramble a bit about something that I think is important to consider in the setting of this story. Through the years, I have been asked many times: why do you live and work out there? I have asked myself that a few times, too. As I have mentioned earlier, West Texas or at least the Pecos River valley and upper Trans-Pecos are not the most awe-inspiring part of the world—just dry, sparse desert. Not much breathtaking scenery, just flat terrain with cactus and mesquite trees. If you wait too long at the stop sign in town a coyote may jump in the back of your truck, hoping to ride off to someplace better. It's a hard place to live and work, and some of the people can be dry and hard too. The desert can make some people mean. And there is the wind—it never stops blowing. There are documented stories of settler women literally going bonkers due to that incessant wind. It doesn't do the men any favors either. There are days that wind just eats at your soul and makes you crazy. You can't think straight when you're out working in a hard wind. Because of the wind, the dust blows all the time and everything you own gets covered in it. The folks out here joke that you could take a brand-new steel pipe and weld both ends shut, cut it open in a year and it will be full of dirt. You can't keep it off you or out of your life. You just have to learn to live with it.

Then there is the summer heat, which is just unbearable. When people say, "Well, it's a dry heat" as if that makes it somehow tamer, it drives me crazy. A raging fire is a dry heat too, but you don't want to be engulfed by it. The desert heat will engulf you—114° Fahrenheit sucks big time no matter how dry the air is. You cannot pick up a tool out of the back of your truck or grab the doorknob to your house without your skin burning to the metal. The heat just beats and beats and beats you down. People have heat strokes out here all the time in the summer and the poor animals, if they live through it, suffer tremendously from the heat and lack of water. In the summer, you find little birds and horned lizards just dead and dried up like animal jerky, lying here and there.

And then there is the cold in the winter. The elevation is 2,400 feet above sea level and there are no trees or mountains to stop the roaring hard cold fronts when they come. When a hard front sets in, the temperatures can go from 70° to 10° in just a few hours. It's painful. Sometimes when this happens, the birds just fall over dead, and their bodies lie sprinkled around under trees and in the pasture. They can't acclimate to the cold fast enough and they just die. Other larger animals succumb to it also, but they usually crawl into a hole somewhere and die. The West Texas desert is a hard place to live and a harder place to make a living of any kind.

The positive reasons I have for living here might seem strange to some. Of course, the main reason is that those resources, the salt water and land, make it the best place in the world for what

I always wanted to do, which was farm shrimp and fish. There is just no better clean salt water anywhere in the world. There is also lots of available land that sells cheaply enough you don't have to be a millionaire to buy some. The resources for aquaculture are for sure the main reason for living here and pursuing my dream, but there are other reasons that are very important also, and make all those issues with the wind, heat, and cold bearable. The sky is one of those reasons. It may sound strange, but something as simple as the sky is a big reason to be here. Most people never give it a second thought. When I was a kid, I looked at the sky and pondered the clouds and the blueness and all that goes with it, but somehow that magical intrigue leaves most of us when we grow up and we live out our adult days and never really look at or even care about the sky.

About the only folks who are interested in the sky other than weathermen on the news stations are airplane pilots, ship captains, and farmers. These folks, especially the farmer, live and die by the sky. The farmer can't change course and avoid the weather like the pilot and the boat captain. The farmer must learn how to study the sky, study what it has to say and what it tries to hide from us. Yes, the sky sometimes tries to trick us, and you must pay attention to it and learn to divine from it its secrets if you're going to try to make a living outside underneath that sky, especially in the desert. And the desert sky is the best sky. The biggest sky. The most amazing sky there is. The only sky that can even come close is the big sky you see when you're at sea. It is sky that goes on forever, big and powerful. I could sit on a shrimp pond levee in north Pecos County and watch a supercell thunderstorm in Artesia, New Mexico, almost 200 miles away, or watch the lightning inside a big storm of weather the same distance to the east. Storms miles away seem to look like they are just down the road. On one hand, it makes you feel like a tiny insignificant speck in the universe and at the same time empowers you to feel like you can do anything. I think that because of the open spaces and big skies there is a freedom here that exists only in the desert southwest—the freedom to think differently from the crowds of people who seem to all act in lockstep with each other. It is freedom to raise your children as you see fit, not how others tell you that you should; freedom to live your life by whatever rules you choose or no rules at all. If you've ever been to Terlingua, Texas, you know what I mean about no rules at all.

The openness also gives you room to create who you are or who you want to be. Through the years, we often traveled east where there are lots of trees and green things, hills, and people. Lots of people. Way too many people, who always seem rushed and angry. Upon returning to West Texas, just about where you cross the Pecos River back into the Trans-Pecos area on Interstate 10, the skies open up and you feel free again. Then you get to the Squaw Teat Mountain in Bakersfield and you feel like you can breathe—you're home again. You are where most of the sun-beaten and heat-worn people will do just about anything they can to help you out. The people out here help each other in order to survive, both physically and mentally. Having a sense of the importance of community, even a community spread out over an enormous geographical area, and helping out your neighbors is paramount to being able to live happily here. There is just something about big, open country with big, giant skies that makes you feel alive and powerful. It makes you feel empowered and capable. It's a sort of magic that gets in you and inspires you. It motivates big ideas, and it makes you believe you can do screwball things like borrow an enormous amount of money to go build a shrimp farm.... Then again, maybe it was that damn wind that made me crazy enough to do that.

Permian Sea Shrimp Company farm.

Paddlewheel aerators providing oxygen to growing animals.

Shrimp harvest pump.

Harvest setup with pump in the water and dewatering box.

Pickers at harvest.

Shrimp coming into the dewatering box.

Patsy taking care of harvest day retail customers.

Washing baskets of muddy picked shrimp.

Feeding a harvest crew.

Permian Sea Shrimp Company packaging and marketing photo.

Judge blocks use of loan information

■ News agency says records are public; businesses say they have trade secrets

By Amy Strahan
Associated Press

A Travis County judge on Wednesday issued a temporary restraining order preventing The Associated Press from using information obtained from the Texas Department of Agriculture regarding a state-financed agricultural loan program.

State District Judge Margaret Cooper issued the order following a request by Permian Sea Shrimp and Seafood Ltd. of Imperial and the First National Bank of Monahans.

The AP obtained the information under the Texas Public Information Act, but the company and the bank argued that the Agriculture Department should not have released the documents because they included trade secrets.

The information released included a written summary supporting the Texas Agricultural Finance Authority Board's approval of a loan guarantee for Permian Sea Shrimp and Seafood.

"The horse was let out of the barn by the TDA; we're just trying to keep it in the corral," said Ben Hathaway, lawyer for the two companies.

Cooper said another hearing should be held to determine whether the information is public and was released properly.

David Donaldson, a lawyer for the AP, argued that the judge's order is an unconstitutional prior restraint on the news media. "Our constitutional principles require freedom of the press," Donaldson said.

Permian Sea Shrimp and Seafood raises shrimp in West Texas

See Judge's, B6

Judge's order on agricultural loans violates freedom of press, AP says

Continued from B1

in Pecos County, about 200 miles northwest of San Antonio.

The Texas Department of Agriculture provided an AP reporter on Wednesday with a copy of the application for a loan guarantee for Permian Sea under a program which the Texas Agricultural Finance Authority guarantees loans to small agricultural businesses.

The authority was created to encourage entrepreneurs to start innovative new businesses in rural Texas by guaranteeing as much as 90 percent of their loans from private lenders.

Generally, the program supports businesses that refine raw materials, such as cotton, that are usually shipped out of state, as well as businesses that grow exotic plants and animals.

Kathy Reed, counsel for the Agriculture Department, said the information released to the AP was public, citing a 1998 letter from the Texas attorney general's office.

"We were relying on a previous attorney general opinion that determined that these credit reports are public information," Reed said.

In that 1998 decision, a request by companies seeking to prevent the Agriculture Department from releasing loan information was denied.

"We believe that the public has a legitimate interest in how and to whom TAFA awards agricultural loans, and we believe that this interest is served by releasing the summary financial information," said the letter signed by Assistant Attorney General Karen Hattaway.

A hearing for a temporary injunction on the matter is expected to be scheduled for next week.

Newspaper article on the Associated Press ordeal.

A yellow shrimp, a genetic rarity found only in pond-grown shrimp.

PART III

CHAPTER 8

PERMIAN SEA SHRIMP COMPANY

B y 1998, I had decided to pull the trigger on starting my own shrimp farm, and I spent quite a few late-night hours putting together a business plan. That was something I had learned from Dan: to live or die by the business plan. My business plan was for a shrimp company that would be better than all the previous farms and would take into account all the things we had learned in the last five years in West Texas. This would be a unique shrimp farm like no other before it. It took a lot of self-reflection and consideration of the performance, both good and bad, of the three other farms we had built and run, as well as what I knew about coastal farms. Once satisfied with the business plan I had written, and happy with the economic forecasts that went with it, I was ready to try to raise the money to start the venture through either investors or a bank loan or both. I was sure there would be no loan available for this project without some kind of loan guarantee. I had no property or capital of my own, so without a loan guarantee there would be no way a bank would get on board with my plan. Of course, even though the bank is who actually gets the guarantee, the borrower must do all the work to obtain the guarantee for the bank. If you want your business to happen, you have to do all the work to make it happen.

Most people are familiar with the Small Business Administration (SBA). They developed the loan guarantee model. They are a government entity that pledges taxpayer dollars for loan guarantees for small businesses and higher risk businesses, which are usually the same thing. They don't give or loan money to the business; they only guarantee to pay the bank back a percentage of the loan if the business should fail. Once a loan guarantee is obtained, the entrepreneur must still find a bank who will loan them the money, and that bank will still ask for 150 percent of the loan amount in collateral even if they know they will have a guarantee of 80 or 90 percent.

As a side note: don't ever believe that banks are here to help you out or assist your business. Banks have one main goal: to preserve the investors' capital (money) at all costs. Loaning out the investors' capital is risky, so they don't do it very often. You keep your money in a bank for years and then need a loan, only to be told that you are wasting your time. It is heartbreaking for people and a bitter pill to swallow. Once you know this and know what truth you are actually dealing with, and if you are ready with a good business plan and have everything in place, you might make your pitch and be successful in getting that loan. But that's only if you understand how the bankers think. And only if you get them a deal that has virtually no risk to the bank.

I did know how bankers think and I knew they were going to want a guarantee. I had a guarantee ready and waiting. This one was through the Texas Department of Agriculture. During my preparation for starting this state-of-the-art shrimp farm I was going to build, I found a program called the Texas Agriculture Finance Authority (TAFA) within the Texas Department of Agriculture.

Rick Perry, a Texas representative at the time, helped bring this program into existence. Then the program was expanded after Perry became the Texas Agriculture Commissioner. TAFA was a creative and helpful program that existed to try to lend a hand to innovative and higher risk agriculture ventures. It was set up to help specialty crops and livestock like pistachios and miniature cows, urban farming, and entrepreneurs trying to do nontraditional, forward-thinking and progressive agriculture. It seemed like a perfect fit for a nontraditional form of agriculture like a West Texas shrimp farm. After filling out stacks and stacks of forms and undergoing two interviews in Austin with their administrators, the TAFA gave me the thumbs up. Next steps: land for the farm and a bank that would do the deal, two very big, loose last ends.

Every night I spent my time burning the midnight oil working on the business plan and financials for this new farm, but during the daylight hours, I traveled all over Pecos County looking for a satisfactory location. The entire north end of Pecos County is suited for shrimp farming in terms of land and water. The Pecos River has laid down reasonably good clay soils in this area that are suitable for building ponds. In addition, most locations have good roads and electricity thanks to the oil industry in the area. The biggest challenge was finding a small parcel of land, less than 1,000 acres. However, land in West Texas is sparse and dry and there is lots and lots of it. Seldom are there ever just a few acres for sale because a few acres is not of much use to anyone in this part of the world. If you want to do anything of any value in this desert, you usually need lots of land to make any endeavor worthwhile. Shrimp farming was a little different, though. It is an intensive type of agriculture that can get a lot of productivity out of a smaller area. I wanted enough land to be a significant player, but in the shrimp aquaculture world, that was just a few hundred acres or so.

It took quite a bit of searching, but eventually I found a few options. I settled on a parcel just east of Imperial that seemed to have been for sale for some time. I had never really noticed it when I was headed out of town to look for property, then one day on my way back into Imperial I saw a very weatherworn, handwritten for sale sign on the fence. I stopped in and met an old widow lady who was anxious to sell the place. Her husband had died some time back, and she was ready to move off the 140-acre farm where she had lived the last fifty years. She wanted to move somewhere where she could be near her grown kids and grandkids. I was careful how I presented the whole thing to her. I had devised a strategy when I started my land search that I was not going to mention anything about shrimp farming. I was just going to propose that I was a young farmer in search of a farm. This was true, although maybe not full disclosure. It was important to be discreet about it all to keep from getting taken advantage of. Shrimp farming had gotten a lot of press and word had been going around that it was going to be huge, and big investors from out of town were going to come in and pay ridiculous prices for land and spend millions of dollars. There were even some folks, mostly across the river in Ward County, buying up land along the Pecos River on speculation that a big shrimp farming boom was coming and they were going to make a windfall. Those guys never sold an acre of land for any shrimp farm, but they did help stir up the locals, which would work against me if I gave too much information. I had to be careful or what would have been $250 an acre land might have turned into $2,500 an acre the moment they knew I was going to build a shrimp farm. So as far as anyone would know I was just looking to farm cotton and hay, and of course in this part of the world that was equivalent to a vow of poverty, so no one ever thought to price gouge.

The old widow mostly just felt sorry for me, I think. After several long interviews to determine my financial and moral worthiness, the old gal ended up giving me a fair price and I gave her my word I would complete the purchase. I then contacted all the neighboring landowners to see if they would also sell so that a large contiguous piece of somewhere around 350 acres could be put together for this new shrimp farm. That part was a lot harder since these people were all absentee landowners. I had to correspond by mail or phone and chase down no fewer than about thirty different people who had all inherited the land through the years. None of them had ever even seen the land, and as long as I wasn't asking for the oil and mineral rights, they were willing to talk. I gave them all the same story: just a young farmer wanting some land to make a farm. I am sure they pitied what they perceived as my poor misguided ambition. All I can say was that it worked, and no one ever asked for more than $265 per acre. I have always found that it pays to keep the details of your business to yourself, especially in this little town of jabbering busybodies.

With the guarantee and the land, I was ready for the bankers. I chose a local bank in Monahans, Texas, at the suggestion of a friend. Since we had gotten so much press and notoriety in the last few years, everyone in every town around Imperial knew what we were doing, and people were genuinely interested. Even bankers seemed interested in it. I doubted that they were very sincere in their interest, but they needed to seem supportive of new industry. They always seem supportive of new industry, especially if someone else is financing it. However, I was ready with my TAFA guarantee, my business plan, and a little 340-acre farm just west of Imperial with the price all negotiated out, complete with electricity and natural gas and right on the main paved road. I even had a few friends and family agree to invest, so there would be a bit of equity money to put into the deal as well. How could they say no?

I got the bank in Monahans interested enough in my project to agree to several meetings, including a presentation to their board. They eventually agreed to accept the guarantee and made the loan. Having a bank finance such a new, innovative, and strange business was significant, and it meant West Texas shrimp farming was seen as a potentially successful enterprise with manageable risk and a short but good track record. This is what all new industry dreams of. It is hard to get to this point, and the local banks rarely feel comfortable enough to get involved in a nontraditional business, but in this case, we proved we were going to work hard, night and day, to make it successful. It seemed that West Texas, and specifically Pecos County, was going be the place where a regular guy from a modest background could own and operate his own shrimp farming business. What an opportunity it could also be for other guys like me. I was on cloud nine.

We began construction in February 1999. First, we needed water. The old farm had a well on it, and I hoped that it would be in good enough shape to use. Then I would need only one additional good well. As luck would have it, the old well was caved in and no good. That meant I would have to drill at least two wells. I had budgeted for this but was not looking forward to it. I was going to have to drill wells with a rotary rig because my old friend Mr. Jeter with the cable tool rig was booked up at the time and could not make it to my farm anytime soon. I felt like the cable tool wells made more water, but the rotary rigs were much faster. It was a tradeoff I would have to accept. But by chance, I heard about a local Fort Stockton fellow who had a cable tool rig and was inexpensive because he just loved drilling wells. It was sort of a hobby but since he also had no other gainful employment, he needed to drill wells to pay his bills. His name was Casey, and while he did not have the greatest reputation in the world for reliability, he had what I wanted—a cable tool rig. He was ecstatic about the opportunity to drill me a well, but he also warned me that he had no hired help to assist him in the drilling and asked if I minded working as his hand for this job. I had no problem with that, now that I had a bit of experience helping folks drill water wells on those other shrimp farms.

He showed up a short time later and went to work. I was sure as fast as he got there and with all his enthusiasm, we would have a well in no time. Boy, was I ever wrong. I am convinced that those cable tool rigs with their constant bang, bang, banging indeed do make the driller a bit screwy. Poor ole Casey was a good fellow, but he was just fraught with a host of personal issues in his life. He would rig up and drill for a few days and just when we started making some distance down the hole, something would happen back in town, and he would disappear for three or four days. Meantime, while he was away, the hole would collapse down deep. When he came back to start another day of drilling, he would have to start all over again to regain what we lost from his absence. It was all he could do to get back to the same depth we had achieved the time before. This went on for weeks: forty feet down, thirty feet caved back in; fifty feet down, twenty-five feet caved back in.

We crept along like this for more than two months and then the fear of the worst kicked in. What if the whole thing caved in, drilling rig and all? Each time that hole caved in, the sides fell into the drilled-out hole. So, if the bit was twenty-five inches in diameter, the hole would be about thirty inches in diameter until the walls collapsed down there inside the well bore, making the hole forty-five inches in diameter. Then all that collapsed the next time, leaving the hole eighty inches in diameter, and so on until we had a giant cavern down there. At the surface, it just looks like a 30-inch hole, but twenty-five or thirty feet down, it's a chasm waiting for a disaster. Casey acknowledged this issue but didn't seem to care. I told you cable tool drillers are nuts! He even lost the bit when it came apart from the cable and fell to the bottom of the hole. He had to fish for it for several days. He finally managed to grab onto it with a tool he fabricated and pulled it out. Then he promptly lit out on a celebratory bender, not to be seen for a week. I was scared. After about ten very long and very nervous weeks we hit our goal of 200 feet. I had all the well casing (the pipe that goes in the hole to stabilize the whole thing) in place. It had all been torch slotted and ready to run. You have to run the casing as soon as you hit your goal depth, or the hole will again cave in, and you can't get the pipe all the way down. Fortunately, Casey had no family issues, parole hearings, or wild Terlingua benders to keep him away, so we immediately ran the casing right to the bottom just like we were supposed to do.

The next procedure was to put gravel in the space between the 16-inch casing and the 30-inch hole. It's called the gravel pack, and the gravel acts as the filter for the water coming into the casing pipe. Normally, this well should have taken about three to four truckloads of gravel to fill the void, but due to all that caving in and the enormous underground cavity we made by Casey's intermittent drilling technique we put twenty-four loads of gravel in that well space before we finally saw gravel at the surface. Twenty-four truckloads! The downside was that cost me a fortune to purchase all that gravel. The upside was that this water well was probably the best water well ever drilled in Pecos County for high volume water flow. That giant cavern was like a miniature ocean under there with no impediment to flow. We put the biggest pump and motor we could fit down that hole, and I pumped an enormous amount of water. It would continue to be an impressive well for years and years. Drilling that well was quite an experience for me. I am sure it took a few years off my life, and I am pretty sure that constant pounding of the rig made me as looney as the driller. Because this cable tool well was taking way longer than I had anticipated, and I knew they would be fast, I went ahead and had West Texas Water Well company drill a second well with their rotary rig while Casey was doing this cable tooling. They had the well drilled in four days. I got lots of water out of that well, and it certainly was much less stressful to complete. I never again had the nerve to drill another well with a cable tool rig and I never suggested anyone else hire Casey either. He was a nice guy, but he was no Cecil Jeter.

Understanding the importance of water wells to shrimp farming in the desert is imperative. The water well is the single most important thing on the farm. Without the water well, there is nothing—no ponds, no shrimp, no business, no nothing. This was why the first person I wanted to talk to when I came out here on that first visit was a water well man.

The well provides that magical water from the Permian Sea and it must keep providing that water day after day, year after year. The well is the first thing you check every day and the last thing you check on before going back to the house. You take care of the water well as if you want it to remember you in its will. You grease it and check the oil every day. You listen for any sounds that are not normal, and you feel the motor to make sure it is not getting too hot. When the storms come, you shut it off so that a lightning bolt will not take it out and ruin the motor. Motors take a long time to get re-wound in this remote area, so we protected the motors at all costs. We also kept a spare motor, pump, and lots of spare parts so that if something serious happened and we had to call a water well service company in to fix it, we had the parts for them to use because wells of this size with this size tubing and pumps are rather rare; there are just not many places that have the kind of water flow that we have here in West Texas. It was a true blessing to have what we called "big water," but we needed water to run all the time; otherwise, the ponds would evaporate or get stagnant waiting several days for water well parts of this size. In the hot, dry, windy desert we simply could not wait days.

It took many years to learn the ins and outs of drilling and maintaining water wells in this saltwater aquifer. We learned that the pumping level of the wells was very high, which was a good thing. It meant that the water in the casing did not pull way down when we turned the pump on. The pump sat at 160 feet and the water sat at thirty feet below the surface when the pump was off. When we kicked on the pump, the water only dropped to about seventy feet inside the casing, leaving the pump covered by ninety feet of water in the casing, meaning it could pump like crazy. This is because the Permian Sea formation is super porous due to the large gravel where the water is located, and the water can just pour into the casing and stay ahead of the pump.

Most wells in other places unfortunately have a pumping level that is often just above the pump, sometimes causing the pump to "pump off," meaning it pumps all the water out faster than the formation can fill the casing. This causes the pump to spin too fast with no load, which damages the pump and motor. Pumping off also means that air gets sucked into your pump, which damages the impellor. At Permian Sea, we did not ever have the problem of pumping off and could easily pump 4,000 gallons per minute in each well with a 100 hp motor, a hell of a lot of water. Because of so much available water, we learned to use the largest casing and pump tubing (the pipe that is connected to the pump inside of which the water flows to the surface) we could afford, which was 16-inch or 18-inch casing and 10-inch tubing.

We also learned that every few weeks, we had to add gravel to the outside of the casing because the pump moved so much water, it pumped some sand out of the formation. This made the hole outside of the casing grow bigger and bigger. If we didn't replace the sand with gravel, we ran the risk of letting the well cave in. We also learned that it would cave in one day no matter what we did. It takes about ten years, but these wells will cave in, and when they do, they ruin the pumps and tubing. The best bet at that point is to just move to a different location and drill a new well. It is just part of the cost of doing business. Some wells will give you more time, and now with the fancy variable frequency drives that allow you to control the motor speed, you can get a lot longer life out of these wells, but salt water and hard work sure take their toll on them eventually.

On occasion, natural gas can enter water wells. Underground formations contain natural gas that can sometimes bubble into the aquifer and possibly find its way into the water wells. It is usually not troubling or dangerous but can cause a flash and a bit of a boom if you're not careful. One time I needed to add some additional casing on one of the wells so I could raise the motor up, so I hired ole crazy Backhoe Ray to come weld it on for me. Backhoe Ray was even a better welder than he was a backhoe operator. When I hired him, I forgot to mention to him that there was a little gas in that particular well. After he had gone to work on it, I suddenly remembered I should warn him to let the welder sparks go down the casing and ignite the gas before he really started working—just sort of flash the gas off at first. I drove up to the well, and there was poor Ray sitting beside the casing with his eyebrows scorched off and his welding helmet blown off behind him. He was shook up and a little bit pissed. He said he put the new casing up on the old casing, put his helmet on over his face and struck the rod on the side. When the sparks fell into the well, it started shaking like an earthquake and then, BOOM, out of the well came a blast, knocking his helmet off and singeing his hair. I was really feeling like it would probably be better for me to play dumb on this issue so I just feigned surprise and told him we should really make a point to check these wells for gas before we weld near them in the future. He went back to work on the casing, but I could tell that he was nervous every time he struck a spark up on that welding rod. I was just glad he was too shaken up to beat the crap out of me.

These water wells and the learning curve of how to drill, maintain, and use them was the backbone of the shrimp farming industry. It is getting harder to find folks to work on big wells or drillers with large enough bits to drill these large wells. During my time shrimp farming we had three good outfits who worked for me from the first farm through the Permian Sea Shrimp Company's existence, and they all served us well the whole time we were in business. They were West Texas Water Well Service, Hartman Water Well, and George Snyder from Van Horn. These three water well companies were paramount to our success while we were in the shrimp business. I really cherished all I learned from them and all the good work they did for us in Pecos County. They got me out of a bind on many occasions, that is for sure.

INNOVATIVE PONDS

When we were ready to build ponds, those same expert dirt movers, Kelly and Terry, who had helped us on that first farm almost seven years before were still here to help me again, thank goodness. This farm would be radically different in several ways from any farm before, both on the coast or out in West Texas. I had gained a lot of experience by now and had seen what worked and what didn't work. My goal was to build a farm that not only operated efficiently and produced shrimp consistently and in large numbers, but also harvested perfectly. I wanted it to reflect the knowledge I had gained, all the setbacks I had endured, and all the successes we had growing shrimp in West Texas. A lot of traditional thinking was going out the window, and so much innovation that some of my shrimp farming friends on the coast really thought I was losing my mind.

As I explained before, shrimp are harvested by draining the water from a pond. As the water drains the shrimp, not wishing to be left stranded high and dry, follow the water as it moves to the harvest structure. Several factors influence the shrimp movement at harvest: the slope of the pond bottom, the shape of the pond, and the harvest pump. We had experimented with steeper and steeper pond slopes with each new farm throughout the years. I wanted a steep slope of more than half a foot per one hundred feet. I also wanted a compound slope where the pond not only slopes toward the harvest end but slopes to the middle as well. While not common for pond construction on the coast, this was an easy enough thing to do with modern lasers and equipment and we had done that before with good success.

What was most radical was that I wanted a very new pond shape—triangle-shaped like a funnel. At harvest, we'd slowly move the bulk of the shrimp across the pond bottom down to the harvest sump by draining the water, similar to how you would scrape crumbs off a kitchen counter into a trash can. On a rectangle table, you always get some crumbs on the outermost edges that you have to re-scrape into the middle so you don't miss the trash can. The same thing happens with shrimp. If the harvest sump or drain is in the middle of the levee on a rectangular pond, you inevitably leave shrimp on the adjacent corners. That means stranded shrimp must be collected manually by physically entering the pond, picking them up one by one into a basket, and carrying the basket out of the pond. It always seemed to me that in a triangle-shaped pond with the harvest sump set in the apex point, then as the shrimp came down during the harvest drawdown, they would funnel into the sump. I observed this in a pond on the Triton Shrimp farm that was sort of triangle shaped. It was built in a strange shape just to fit in the last available bit of land, but it harvested the most efficiently of all the ponds on that farm. The scene of that harvest in that triangle-shaped pond never left my mind, and I was soon convinced that modern shrimp pond construction and design was all wrong.

The real dilemma became how to efficiently build triangle-shaped ponds on square land. Land comes in all shapes, but it is mostly laid out in square and rectangular plots. That's just how surveys are done. What we finally did, after a lot of thought, was settle on a configuration of six 3.5-acre ponds built in the shape of a square wheel with the spokes making each pond triangle shaped. It's hard to visualize, but think of a rectangular pizza that you cut into triangles and that might help a bit. It was very radical for pond design. The pond bottom slopes would be falling toward the end point of the triangles at the hub of the square wheel. Each pond had a steep bottom slope creating a great big funnel. This revolutionary new design, coupled with placing all the harvest equipment in the same general middle hub area for all ponds, would mean minimal set up and breakdown time—harvest would be simple and efficient. It was progressive, radical, and super cool because it was a new kind of grower-friendly farm, designed from experience of work in the trenches by a real shrimp farmer, not an academician or civil engineer. A farm mapped out by the scars from years of experience, both good and bad.

In addition to the radical pond designs, we also implemented the most radical harvest method ever used in the shrimp industry anywhere. The Pecos valley is not blessed by much elevation difference. In other words, it is pretty darn flat, which poses a challenge for draining water. If you want to drain a pond, then you must dig a drainpipe lower than the bottom of the pond. Then you must dig a ditch lower than the drainpipe to carry the water away to some area that is, again, lower than the ditch and large enough to hold all the water from the pond. Pretty soon you have dug yourself to China trying to get rid of water. On the previous farms, it was problematic but we found work-arounds. The Triton Aquaculture farm had some old gravel pits that were dug pretty deep, and we diverted harvest water into those. At Regal Farms there were enough acres of open land and just enough elevation fall over the entire property to run the water out to the far side of the property and away from the ponds.

The property for the Permian Sea Shrimp Company was just flat. This property had been a hay farm for fifty years and the land was flat as it could be, so there was no way to economically drain a pond because there was no place lower than any other. This indeed was a troubling issue. Then my friend Kelly, the dirt guru who had helped me design and build ponds from the very start and was going to do the construction on this farm also, had an idea. He imagined a system that would pump both water and shrimp out of the ponds at the same time. The water would be diverted in one direction to either be stored for possible future use (in another pond or a holding basin) or fully released into a field, and the shrimp would be separated and collected in a different area—all at the same time, no drains needed. Shrimp had never been harvested this way before, but it was a promising theory.

We went to work on this idea—right or wrong, smart or stupid—both deciding this was for sure the route we were going to take, pumping shrimp and water all at the same time. While we were working on this wild new harvest idea that no one had ever done before, we rolled the dice and started construction on ponds that had no drain system built into them. The ponds would have to be pumped out to get the shrimp out at harvest time, and that harvest time was a mere nine months away and coming up fast. This was the ultimate commitment to an idea, especially considering that we knew no one in the world who had worked with this pond configuration, nor was there a pump in existence that could do the job—yet. We would have to design and build the pump and figure out this whole new harvest system all while building the ponds and getting the first crop in.

The people I knew in the Texas coastal shrimp industry thought I was crazy when I told them of our idea. They were convinced that the dry, hot desert air had gotten to me. Undeterred, Kelly and I sat up many late-night hours designing the harvest pump because during the day we had to build ponds, drill wells, and run piping and electrical lines. We were on a very tight timeline, similar to what we faced a few years previous with Regal. We had about three to four months to build all this and have ponds ready to stock with larval shrimp. Kelly, his hand (brother), Little Manuel, and I would build and equip ponds for twelve hours a day, and then Kelly and I would meet up and work on the harvest pump design at night. I had full confidence that a large centrifugal pump like the ones used to irrigate fields out of lakes, creeks, and ditches could move a lot of water out of a pond. Centrifugal pumps can do that with little effort and little horsepower. The trick was getting the shrimp separated out of the water at the same time—shrimp don't like a harsh current. They will jump away from it the moment they feel the water pulling them. You have to finesse shrimp out of a pond. From experience I knew I could get the shrimp to follow the water to the deep end as the pond level slowly went down, but once they were concentrated in the sump area, how would we get them to come into the pump, and how would we keep from shredding them to pieces by the impellor of the large water pump?

To solve the first dilemma, the problem of an intense intake suction, we designed a very large 48-inch diameter and eight-foot-long intake pipe on the front of the large irrigation pump. This would provide a very gentle suction regardless of how much water the pump was moving, which was going to be about 7,000 gallons a minute. That is a lot of water, but the large opening would allow a large volume of water to enter the pump intake slowly, and the shrimp would be fooled by the low current and follow the mild current right into the harvest pump system. Next, we designed a giant filter screen inside the pump intake, which would allow the pond water to pass through, but the screen would have gaps too small to let the shrimp pass through. The screen was shaped like a giant funnel inside the 48-inch-intake pipe. While the pond water passed through the screen and into and out the large centrifugal pump, the screen would funnel the shrimp to a hydraulic fish pump inside the housing of the whole apparatus. It is a lot like using a kitchen strainer to separate water and noodles, only in this case, at the center of the strainer would be a pump that would constantly remove the noodles from the strainer so you could keep adding more noodles to it. It was a very innovative idea. At the outlet of the whole pump contraption were two hoses. One was a 16-inch hose that carried the pond water to wherever we would designate it to go, like a ditch or another empty pond. The other was a 6-inch hose coming from the shrimp pump and carrying the shrimp, and just enough pond water to keep the shrimp moving, to a large metal container with an angled screened bottom. We called this box "Big Blue" and we it made from a 55-galllon drum, some expanded metal, and an old trailer to hold the entire contraption. The shrimp, perfectly intact, alive, and kicking, would enter this dewatering box and the water would fall down through the screen bottom and shrimp would slide down the angled bottom to a manually operated door that when opened by a harvest hand would release the shrimp, alive and hopping, into the shrimp baskets.

A large tractor would power all this magic by using its power take-off to run the big centrifugal water pump and the tractor's hydraulic system to run the shrimp pump. Once we were happy and confident in our pump design, we sent the drawings to an agriculture pump manufacturer along with a big deposit of money and a small timeline. They had to have this crazy thing built by no later than the first day of September, only a few months down the road. We had the pond construction finished by May with only one real hitch. An unusual hazard occurred on the farm while we did all that scraping and pond building—something strange that none of us expected.

Anytime you're digging around in the dirt, you have to be aware of things in the ground like pipelines, waterlines, and electrical lines. For a while now, Texas has used the 811 protocols wherein you call 811 and they come out and tell you if it is safe to dig in the area.

The information the folks from 811 have is only as good as the historical data that they have on record for what is buried in your area, however. In the Pecos River valley people have been burying stuff for years. There are old pipelines and oil wells that date back to the 1920s, long before anyone kept any records and long forgotten by even the crustiest old-timers. About three ponds into the build, one of the scrapers hit something big and immovable. That scraper was moving about 15 mph and this thing stopped it dead in its tracks, just about sending the operator through the windshield. We moved the equipment out of the way and did some hand digging in the area and found that we did indeed hit an old, abandoned oil well.

Since I had many maps of the farm and the whole local area, I did a little investigation. A very old map from the 1930s showed that there had been a well location in this very spot, but all the newer ones only showed this area to be farmland. We called Backhoe Ray to dig out around the old well, which he was happy to do for us. Then, we pondered what to do about it. I asked some of the oil men that I knew around Imperial what they thought I should do about it and they told me that I had two options: call the Railroad Commission and ask them what to do, or just leave it for someone else to deal with later. It was right in the middle of what was going to be a pond and I needed every pond to be built and producing in order to make the economics work so just leaving it was not an option for us. I found the local Railroad Commission field tech for the area and asked him what to do. He told me that it should be plugged or capped, but since it was so old, and the Railroad Commission had no record of its existence, getting permission and all the paperwork to plug it could take years and years. He told me with a sly smile to just weld a cap on it and forget about it. "We didn't know it was there for the last seventy years and as far as I'm concerned it doesn't exist," he said. "But I never told you that," wink, wink. It seemed this was my only real, practical choice, and it came straight out of the mouth of someone from the Railroad Commission!

We had already dug out around the casing, so we made a plan to cut that casing off about six feet below the level of where the finished pond bottom would be, fill it up with cement, weld a cap on it, and then cover it back up with dirt so it could live forever under the bottom of that pond. We got everything ready, but no one had the nerve to use a cutting torch to cut the casing. It could be full of natural gas that might have seeped into the casing through the years, or hell, it could be open to a gas formation and the flame of the torch could set off an inferno that would take Red Adair's fire and rescue team to put out. While we were standing around debating the whole thing like a bunch of freshmen boys at the school dance trying to get the nerve to ask the girls to dance, ole Casey the cable tool driller said, "I'll cut that casing off, I ain't afraid of it."

Like I said before, I always thought the pounding of the cable tool rig made the well drillers crazy, and now I was sure of it, and I was damn grateful for it too. Casey took his torch to the large hole in the ground where the casing was sticking up and he went to work. We all stood by, hoping for the best—and by "stood by" I really mean about 250 yards away. When he started cutting there was the smallest little pop from a tiny bit of casing gas igniting from the flame when the torch first cut through the metal, but then that was it. No drama, no fire or explosion, nothing. He went on and cut off the casing smooth and straight. All the nervous energy that had built up in anticipation of imminent doom suddenly left us and we got busy pouring all the cement we had down that hole. We followed it up by shoving all the plastic and burlap bags we could find into whatever room was left inside that pipe, along with one canvas tarp just for good measure. Casey then took a thick piece of metal and welded it over the end of the pipe for a good solid and permanent cap. We were back in business. We finished the pond construction with no more serious setbacks or roadblocks from that point on, and that was good, because I had had about all the nerve-wracking situations I could stand.

Once the farm was finally ready for shrimp, we stocked the almost seventy acres of brand-new ponds with larval shrimp. We picked up the little shrimp from the hatchery in South Texas called Harlingen Shrimp Farms (HSF). We could not just order the shrimp, drive down, and pick them up—it was quite a ceremonious event to procure larval shrimp in Texas. By this time in shrimp farming history, there were a number of farms along the Texas coast, in addition to the ones out in West Texas. Demand for the post-larval (PL) shrimp was high and the coastal farms not only needed lots of PL's, they all wanted them in the month of April if they could get them then. It was a very competitive environment around shrimp stocking time. To make all this work and keep the peace, HSF would hold a larval shrimp lottery at their headquarters down in Bayview, near Harlingen, in early February. Shrimp farmers from up and down the coast and a few other places would gather in a large meeting room. There would be tickets shuffled up in a bucket that were ready for each farmer to draw out. On each ticket was a predetermined number of so many thousand larval shrimp and dates ranging from April 1 to June 15. For me it was no pressure. I could not stock in April because it is still too cold out west then. If I were to draw an April ticket, I knew I would have no trouble trading it to a coastal farmer for a May ticket. The coastal guys did not want May tickets and none of us wanted June tickets.

I enjoyed the leverage I had knowing that every coastal farm would want to trade for my ticket if I drew one or more in April. They needed April tickets, and they would compete for my favor in order to get a trade. I was a popular guy. They would always be so nice to me and talk to me like I was an old lost friend. They would offer to take me to lunch, or to share good deals on farm equipment like aerators and pumps. See, there were eight or ten of them and only one of me with my preference for May. The whole thing was a lot of fun and camaraderie; it was the one time of year the whole shrimp farming industry was in one room. Sort of a shrimp farming "think tank" for a day. Once spring came, we would drive back down and pick up the baby shrimp based on the dates from the lottery tickets then drive them back the twelve hours to West Texas. This drive had to be done quickly, flawlessly, and at night when the air was cool. The trip was hard on the little baby shrimp as well as us drivers, but the hatchery always produced really good healthy shrimp that could take the long trip and then acclimate to the ponds with good success.

Harlingen Shrimp Farms was an integral part of the shrimp farming industry not only in Texas but also all over the country during the peak of shrimp farming in the United States. Formerly known as Laguna Madre Shrimp Farms, named for Laguna Madre Bay where they got their water, it was a huge farm and the biggest and best hatchery in the state of Texas at the time. They provided shrimp to every shrimp farm in Texas and to farms in South Carolina, Florida, and other states as well. That the hatchery was started by a couple of partners from Midland, Texas, some few years before always interested me. Midland is just down the road from Imperial, and I have always found it so ironic that some folks from West Texas who were interested in shrimp farming went all the way down to South Texas to start a shrimp farm when what I think is arguably the best shrimp farming region in the world was right in their backyards!

To be fair, they did start their farm a bit before our heyday in the area, so perhaps they just didn't have enough information about the available resources here, or maybe they just didn't understand how awesome the possibilities were for shrimp farming right here in West Texas. I just always found it perplexing. Harlingen Shrimp Farms with their Midland investors did have a fine facility down south and always produced excellent larval shrimp for stocking into ponds. One advantage to stocking later in the spring was that HSF could provide us with larval shrimp that were a bit older and bigger. In April, with all the large orders, they got the larval shrimp out the door as fast as possible. In May, things slowed down and they had more time and could hold them in the larval tanks a bit longer. This was great for us: since the older babies traveled the long distance better, they could acclimate to our water. They also grew better when started out older. Once those babies were here, they loved it and did us proud each year. I hope those Midland guys were proud too, you know, for the hometown boys.

During that first summer while the shrimp were growing up in the ponds, we had a bit of a local stir concerning those fellows across the river that I mentioned before, who had bought up all that river land anticipating a giant windfall when the industry took off like a prairie fire. Well, the industry was just slowly moving along more like a Boy Scout campfire than a prairie fire. These fellows just could not wait for it to evolve on its own and decided to force something to happen. It could have been a legal disaster for them and a black eye to the whole area, but it thankfully just turned out to be a laughable blunder by a couple of guys who apparently specialize in blunders. First off, you had to understand these guys. They may have had other partners, but I only ever knew of these two knuckleheads, and they were the main agitators. Early on, I had met one of them at the PCWID No. 3 demonstration farm when I was new in town. He drove up, jumped out of his truck, and introduced himself as a very important mover and shaker in the area. He told me he could put me in touch with all the right people and dealmakers. Of course, he would need to be paid for all this hard work of connecting people. Apparently, he was not clever enough to do much of anything himself or even be an integral part of a deal, he just wanted to be paid to make stuff happen and for other people to do the work while he got the credit. I was so unimpressed with him on that first visit that I totally forgot about the guy. He and another man kept popping up, however, doing ridiculous things like buying all that land along the river and trying to promote an industry that they knew absolutely nothing about. But it didn't bother me because they also promoted lots of other things; other deals and projects they wanted to see happen on their side of the river and in their county. Mostly I think they just wanted to get credit for things other people were doing to make things happen. They never knew it, but they were notorious in the area. You could not mention their names in public without seeing eyes roll.

Since nothing was moving fast enough in the shrimp farming industry to meet their satisfaction, and they were sitting on all this idle land that they were sure would have been acres and acres of shrimp farms by now, they decided to shake some trees. What fell out were convicted criminals, and we almost had a real mess. Somehow, these fellas found a couple of people posing as international financial experts who said they wanted to start to develop miles of river land into shrimp farms. They were going to bring in an army of Chinese shrimp farming experts to run these farms. They told everyone how well connected they were politically and flaunted an impressive resume of big projects they had participated in. It started a buzz around West Texas. I checked into their proposed plans just out of curiosity and became suspicious fast. It just didn't add up. The promotional materials had all the buzzwords like jobs and economic growth, but I could find no real information on an actual shrimp farm: no designs, operations, or marketing plans—nothing.

The one thing we learned for sure was that they were trying to get the state to pledge money for their project and were in Austin constantly working the politicians. That part was infuriating. Those of us who were in this industry had put up our own fortunes and risked our own money or were borrowing money and risking a lifetime of debt and ruin, and these yahoos wanted a government handout. After a bit of sleuthing, however, a friend of ours found out that these international financial gurus were convicted felons and had records of bank fraud, wire fraud, and conspiracy. Upon further investigation by several now very interested citizens, it turned out what these men were planning to do was implement a giant immigration scam, with Chinese nationals. By saying they were aquaculture specialists, they could obtain visas for the immigrants since there were not very many Americans who understood this kind of work; they hoped the government bureaucrats would agree that we needed to import that knowledge. The plan was to get a big payment from each Chinese person or family who wanted to come to America. They would bring them to West Texas and put them to work on the fake shrimp farm for a short period. That was the story to cover up their plan to make the immigrants disappear into society, making room for the next paying Chinese national who wanted a chance at a life in America. There was potential for them to make a lot of money.

I am sure the two local men had no idea they were being conned. They were not smart enough to see through the scam, and they were so intoxicated by the thought that everyone was going to fawn all over them for bringing this economic revolution to the area that they paid no attention to the glaring sketchiness of their so-called partners. Fortunately, some very smart and attentive local people sniffed this out before it got very far and the whole thing fell apart like a bad pair of Chinese tennis shoes. Once those two locals realized what they had almost gotten themselves into, they were smart enough to put a lot of distance between them and these outside players. Soon the whole thing evaporated into the dry desert air, and no one said another word about it. I heard from some of their friends that the thought of what these two local yahoos almost got themselves into scared those two dudes so badly, they put all their river land up for sale and never breathed a word about the shrimp business again. Thank God.

Now that the schemers were out of the way, we doers could focus on our businesses. As if it wasn't enough work taking care of the growing shrimp, we also had to spend a considerable time that summer dealing with the Texas Parks and Wildlife Department (TP&WL) on some ridiculous regulations that all desert shrimp farms wanted eliminated. As I mentioned early on, there was the rule that stated a shrimp farmer had to notify a game warden each time there was a harvest so they could send someone to make sure your shrimp would not somehow get into the Gulf of Mexico. It was silly, and the game wardens eventually told us to leave them alone. We had, however, another rule that applied to all shrimp farms in Texas without exception: that every year each farm had to submit a hurricane evacuation plan to TP&WL. This plan had to show how you were going to carry out the destruction of all your shrimp ahead of a hurricane in the event of a storm surge or flooding rains, which could wash the shrimp into the waters of the Gulf of Mexico. Remember that by this time all shrimp farms in Texas were growing the Pacific white shrimp, which is exotic to the Gulf of Mexico and thus can't be allowed to escape into said waters.

Along with this evacuation plan is a legally binding agreement that you will carry out this plan and kill your shrimp when an imminent hurricane was approaching the coast. It sounds crazy for farms way out in West Texas to have to follow this regulation, and we asked many times to be exempted from these rules and some other rules concerning disease testing and monitoring, which were all there in an effort to protect the Gulf of Mexico. We were 800 miles from the Gulf of Mexico, and we certainly were not going to kill our shrimp when a storm approached the coast so far away, but it was our legal obligation. We could get charged with noncompliance and lose our permits and licenses. In addition, we just didn't want to have to blatantly disregard the laws. It was a sound rule for farms on the coast, but one that didn't apply to us, and we wanted that made clear and codified in the rules. For years, TP&WL would not budge, then a very fortuitous thing happened. Ernest Angelo, a prominent businessman, sportsman, and super good guy from Midland, Texas, was appointed commissioner to the TP&WL Department. Finally, we could appeal to a fellow West Texan. After a good bit of wrangling, we set up a meeting with Commissioner Angelo and some of the other commissioners as well as biologists and technical folks from the department. At that meeting we laid out our request and the facts and conditions to support it. Commissioner Angelo was flabbergasted and a bit embarrassed that we were still being forced to comply with a rule that was obviously intended for the coastline. There are no hurricanes in West Texas and even if the shrimp got away from a pond out here it would have to travel 800 miles to get to the coast—an impossible feat. Fortunately for us, the commissioner was well respected and had a lot of influence. He finally convinced the rest of the TP&WL Department to exempt us from this and any other rules that were obviously not applicable to West Texas shrimp farming. It had been an exhausting summer.

The shrimp grew nicely that first season, and by the end of September they were all about twenty-two to twenty-four grams. This meant they would produce about twenty-six to thirty tails to the pound, large shrimp, the size we wanted to harvest. If the shrimp grew too slowly or the spring was too cool, the shrimp would turn out smaller and bring a lower price. On the other hand, if the shrimp weighed more than twenty-five grams, we would most likely have low survival in the ponds. The shrimp would have plenty of room to grow big, but the ponds would yield lower total pounds of shrimp. We needed a balance between many pounds and a decent-sized shrimp that would bring a higher price.

The shrimp harvest season in West Texas starts in early October after the weather cools down a bit, but well before the first cold weather. This harvest window in the Trans-Pecos is very tight, and if you wait too long you could have cold, dead shrimp. I was nervous that first season. I had hundreds of thousands of pounds of shrimp swimming around in seventy acres of ponds and no way to get them out until that crazy new harvest pump arrived. Even when the shrimp pump finally arrived in late September, there was still no way of knowing if it would work or not. No one had ever harvested shrimp like this before, and our design was based on our experience with shrimp harvesting and moving water with pumps, but it had all only been proposed on paper. It was probably one of the riskiest things I ever did in my whole career. Hundreds of thousands of dollars all on the line on the assumption that our new contraption would do what it was supposed to do. The giant monstrosity was not even painted when we hooked it up to the tractor and backed it into the first pond to be harvested that year. That pump was just a big, black hunk of hope made of steel. I crossed my fingers, put on a confident smile, and cranked it up. Inside I was scared as hell. If it didn't work, I was ruined. There would be no way to get the shrimp out of the ponds, and all would be lost. I would be bankrupt, and I would not be able to look anyone in the shrimp business in the face again if this failed. They already thought I was crazy for trying it in the first place.

The good Lord was with us, and the whole process worked beautifully, just as we had designed it. The shrimp filtered into the hydraulic pump, out to the dewatering box, and into the baskets and ice. In the meantime, the pond water was diverted to an old irrigation ditch and down to a holding pond we had built about half a mile away. The whole design was a thing of beauty and probably the biggest advancement in pond harvesting of shrimp or fish in one hundred years. Once word got out on how elegantly simple harvesting our shrimp was, all the folks who thought I was crazy were asking me if they could have one built. Being a pioneer sometimes has its glory if you can survive the arrows.

That first pond we harvested on the Permian Sea Shrimp Company farm was a real milestone for me. No longer was I working more than eighty hours a week for someone else to make the money. It was now my farm, and the hard work was paying off for me and my family. Seeing those first shrimp come out of that pond was one of the happiest times of my life. It was also a bit chaotic because all the other farms up to that point had been owned by outsiders from faraway places. They were not particularly friendly to any of the locals, nor did they support or participate in the local community. After the initial novelty of it all wore off, the locals were not that interested in the goings-on of the shrimp farms. By this time, however, I had been working hard around Imperial for about seven years, and for good or bad, I was starting to be seen as kind of a local, hometown boy. When I went out on my own and started my own farm, I got a lot of support from the people in the area here. They sort of felt like a part of it since a local person, one of them, now owned a shrimp farm.

On that first harvest everyone in the local area showed up to see the first shrimp come out of the water on the new farm. It was a big ole party that night. There was food and drink and lots of fun, but while most folks were whooping it up, some of us still had to work and get the shrimp out. Once harvest started, it couldn't stop. I would check on the tractor and pump and then check the pond, then I would go see how the people were doing. Back and forth, back and forth. The folks were having a pretty good time and after a few hours many were also getting a little bit inebriated. They were not having to do any of the work, so all they had to do was party and share the joy. At one point, while I was checking on the pond status, the local constable came to me discreetly and said, "This is a great party, but you better go check on things over there, they are all just sharing the love and the shrimp too!" Puzzled, I eased over to the crowd who never stopped patting me on the back and congratulating me on the great harvest. While I was amongst the party, I could see that my employees were just filling up anyone's and everyone's ice chests with free shrimp. Everyone was just so happy and supportive and felt a part of it all that I guess they forgot that I had to *sell* this shrimp to make money. I was not doing all this just for the fun of it. They all had just got caught up in the moment—but that moment was short-lived. While I was not angry or terribly concerned, and I realized that no one really meant any harm, I did make it perfectly clear that the last of the free shrimp had been distributed and the rest were going to the shrimp buyer for money. Imagine that. Overall, that was a great harvest and the first of many good harvests to come for the Permian Sea Shrimp Company.

KIDS, PICKERS, AND FOOD

We got most of our labor from the local high school in Imperial. Little Manuel was always my main hand, and my son Kolton would always work hard for me, but we often needed more help. There were many good, hardworking teenage kids in town who would work for us during the summers and in the fall for harvest. Each fall they would go to school during the day, go home and sleep a little, then meet us at about midnight to work the harvest, typically until sunrise. Once all the shrimp were loaded, those kids would clean up and head to high school. I had several good kids work for me, but it was the high school girls like Crystal and Meg who were the most reliable and hardworking of them all. They worked circles around the guys! I laugh when I think about all those kids, now grown up and telling their children about when they were shrimpin' in the desert as teenagers. I bet their kids don't believe a word of it. Those young folks were great shrimp farm hands, and I will always appreciate their help.

We would also get day help from any other locals looking for spending cash. It certainly was not easy money, but it was a way for some folks to earn money to feed their vices for sure. On any given harvest day we needed folks to help with a variety of chores, but mostly we needed those all-important pickers. Even though we had improved the ponds and pond harvesting, and the shrimp usually harvested well, it was still necessary to have folks available to go out into the newly drained pond and pick up any shrimp that refused to follow the water and go out the pump. Same story as before: put 'em in a basket, and when the basket is full carry it out, wash all the black pond mud off the shrimp, and then pour them into the tote of ice. Usually, the shrimp would harvest clean, and the picking job just took an hour or so of getting any last stragglers. Other times, usually due to very cold weather or an unfavorable wind, the shrimp would strand all over the pond and then the work was on. Due to the improved pond construction on this farm, I think our record for picking was only around 2,000 pounds of shrimp. That still took a lot of time to pick up, but it was much easier than some of the picks from years earlier. Due to the muddy and dirty nature of the picking job we still had to hire some rather unsavory characters. I learned early on that if I paid these folks as soon as the pond was finished, I would not see them again for a while, so we generally hired two crews. One crew would work about three or four ponds, get paid, and promptly go on a bender. The next crew would come in for the next three or four ponds, get paid, then it was their turn. By the time the second crew was leaving with a pocket full of cash, the first crew had run out of party favors, sobered up, and was in need of money for the next round of partying. They would show back up and be ready for work and the cycle would continue. This crew rotation worked pretty well during each harvest season for all the years we did it. We just had to learn how to work with all sorts of folks.

The best way to work with all kinds of folks is to feed them while they work. Everyone loves food, and in agriculture it is almost a requirement to feed your hands. When you ask people to work crazy hours doing crazy things, you need to feed them. It is your duty as the boss. This is why ranchers feed cowboys who are working cattle, and farmers feed harvesters picking crops. I bet even cannabis farmers feed their harvesters . . . well, maybe that would get too expensive. Anyway, we always fed our workers well on harvest days. One of the best things for breakfast that my wife came up with was tamale, egg, and cheese burritos. She cooked the eggs, cut the tamales into them, added cheese, and rolled it all up into a flour tortilla. Everyone loved these so much she could hardly make enough of them. A burrito makes handy food for shrimp harvesting, which is a good thing because just about breakfast time is almost always the main shrimp rush through the harvest pump. You can't stop working when shrimp are coming fast and furious, so you have to eat on the fly and a burrito is perfect for that. Patsy would get up at around 3 a.m. and prepare breakfast for the workers. She made lots of different breakfast foods, but the tamale and egg burritos were what everyone asked for time and again.

For lunch, we would usually grill up some venison or goat. I often would get Big Manuel to cook for us on harvest days. That guy could cook the best *cabrito* (goat) I ever tasted. Sometimes he would bring massive amounts of green enchiladas for our lunch. No one ever left hungry, that's for sure. I also know that some of those down-and-out pickers that we had working for us really appreciated the food. It might have been the only real food they had that week, and we were happy to provide it for them. Often A. C. and Judy Stephenson would bring food for all of us. They always made fancy food that most of us had never heard of before, but everyone enjoyed it whether we could pronounce it or not. We also always kept a gigantic food-service-size pot of very strong coffee going. On cold harvest mornings I would add liberal amounts of Irish cream and maybe a shot of Jameson to that pot (don't tell OSHA). Food was such an important part of harvest, when I think back on past harvests it's not the shrimp I remember, it's the food that first comes to mind, then the folks. They were all good folks, from the pickers to the buyers. There is nothing on earth like watching the sunrise with that warm Irish coffee in your hands on a cold October West Texas early morning shrimp harvest.

One interesting thing I do want to mention is the yellow shrimp. In any particular pond harvest of twenty to thirty thousand pounds of shrimp, you will get maybe ten yellow-colored shrimp. I mean fluorescent lemon yellow. Science has yet to figure out what this phenomenon is all about. It is mostly considered a genetic expression like albinism, but no one knows for sure. It is not seen in wild-caught shrimp probably because a bright yellow shrimp would be quite an easy target for predators. We always found them interesting and collected them as we harvested. Word was that the Asian shrimp farmers in South Texas looked on them as lucky. We just thought they were cool and watched for them like kids on an Easter egg hunt.

For the next three years we consistently produced 6,000 to 8,000 pounds per acre and got $2.75 to $3.50 per pound of whole shrimp at the farm. Every fall the coastal buyers were happy to send out trucks full of totes and ice, load up the harvest, and write us a check right there on the spot. The truck would head to the coast with the shrimp, and I would head to town to deposit that check and then come back and set up the next pond for a harvest the next day. It was the same process until all the ponds had been harvested. To make extra money, we would advertise the harvest days in the local area newspapers and on local radio. Fresh shrimp retailed for $5.00 per pound right out of the ponds to whoever showed up on harvest morning. There were mornings we sold $10,000 to $15,000 worth of shrimp to the public right there on the pond levee. Folks just could not get enough. Anyone who ever tasted them knows what I mean—they were the best shrimp ever produced anywhere. The coastal buyers and processing plants loved our shrimp too, and they would all compete to get our shrimp. We had great working relationships with those folks on the coast, and together we did a lot of business and had a lot of laughs. One laugh almost got us into big trouble with one processing plant in the middle of a harvest season one year.

First, you need to understand how the processing part works. After the truck full of shrimp leaves the farm, it arrives at the processing plant and the first order of business is the de-heading process. Although we live in a very modern time and most labor-intensive jobs have been replaced by machinery that can do those mundane jobs for us, shrimp heading is another matter—it is still done by hand. A tote of iced shrimp is removed from the truck with a forklift, brought into the processing plant, and dumped out onto a giant heading table. All around this table are people, usually women, who start grabbing the shrimp with both hands and de-heading them. They dispose of the heads in a bucket and the tails either in other buckets or on a conveyor that takes the tails to be graded according to size by a machine. There may be a hundred or more folks around that table heading shrimp dumped out 1,500 pounds at a time. The faster the people work, the more they get paid, so they work fast.

What got me into trouble is that one year just before a harvest, we found a very large rattlesnake near the area we were going to be working. We killed the snake with a grubbing hoe and tossed it aside. Out in Imperial we see rattlesnakes almost every day, so we thought nothing of it. We kill them when they are in our way, and we leave them alone when they are not. Later that day while harvest was fully underway, someone had made a comment about the size of the snake and wondered if anyone on the coast had ever seen a big old West Texas rattlesnake. That gave me a thought that made me snicker a bit. I pondered how funny it might be if I put that dead snake in a tote with the shrimp and ice and sent it with the load to see what those folks thought of it. Seemed innocent enough to me at the time, so into a tote the snake went. We finished the harvest and sent the truck off with a full load.

After a few hours had passed, I had actually forgotten about the snake in the tote. That is, until I got a phone call about a day later. There was a very loud and angry voice on the other end of that phone. It was the voice of the man who owned the processing plant and who also used to be my friend up to that point. He was not a happy guy. Seems the de-heading was all going nicely until one tote got dumped out in the middle of the table and in it was a very large West Texas rattlesnake. That room exploded with women screaming and dashing for their lives. They crashed through doors and windows, destroying anything in their way to get out of there. Apparently, no one, in the intense panic, took the time to realize that the snake was headless and dead. A snake is a snake, and those ladies wanted no part of a snake. He told me that all nerves were shot after that, and the workday was shot too. He had to plead with them and offer more money just to get them to return to work the next day. He was pissed. I was embarrassed. I guess I had not thought the whole joke through and what the ramifications of it might be. I apologized repeatedly, and he made me swear that I would never pull a prank again or we were done doing business. I never put another snake in a tote ever again, but I still think it was funny. So shoot me.

THE ASSOCIATED PRESS ORDEAL

A couple of years after the formation of the Permian Sea Shrimp Company circa 2000, we got caught up in a bizarre political situation, the sort of thing you can't even make up. We were used to being seen as an oddity and a thing of wonder. I knew it was unusual to be farming shrimp in the desert and from time to time had to deal with some special situations like when the bank inspectors from Austin came to visit the farm to make sure the bank was telling the truth about loaning money to a shrimp farm way out here in the desert. They were a bunch of nerdy little men in expensive suits trying as hard as they could not to get dirt on their ostrich skin boots. They weren't at the farm for five minutes before they had seen enough and were satisfied that it was for real and ready to go back to civilization. Or the time I was selling shrimp at a Fort Worth farmers market and a passerby scolded me and called me a liar because if shrimp farming in the desert was real, she would have seen it on the *Texas Country Reporter*.

I did make it onto the *Texas Country Reporter* some years later, but before that we got some public notoriety that was a lot less enjoyable. As I mentioned before, Rick Perry had helped set up the TAFA some years earlier when he was a member of the Texas Legislature, and later the Commissioner of Agriculture. In 2000, he was the governor of Texas, and he was running for another term—he had assumed office when George Bush resigned to run for president—and his opposition was out for blood. I do not know who actually initiated the opposition research, but it was the Associated Press (AP) that did the all the digging. Apparently, they were looking for whatever dirt they could find on Rick Perry and were examining everything he had done from his childhood through elected office. When they discovered TAFA, it must have seemed like they found the smoking gun. Due to the risky nature of the loan portfolio that TAFA guaranteed, there were a sizable number of failures that had gotten loan guarantees through the program. Failures are good fodder for political opponents to pounce on. Never mind that a lot of good, clever folks were trying to change agriculture and the Texas economy in general, Perry's opposition's plan seemed to be to spin it to look like taxpayer money was just thrown away on crazy, untested ideas.

If someone had a new and novel idea that had never been done before, detractors assume there must have been some illegal scam to get a loan for it, or maybe someone came up with crazy ideas to scam the taxpayers out of money, and no doubt Rick Perry was behind it all. He had to have hatched this whole TAFA thing up to enrich his good ole boy Texas friends, right? They were sure they would find his fingerprints on something dirty. First, the loan recipients don't get a dime of the TAFA guarantee (taxpayer) money, the banks get the money if the enterprise fails. Second, Rick Perry might have approved or even came up with the idea for TAFA, but he was never involved in any approvals or loan hearings or anything. I went to Austin several times for meetings, and I was on the phone with lots of folks from TAFA, but I never talked to Rick Perry or Susan Combs, the Commissioner of Agriculture at the time. Instead of hurting Rick Perry, what the Associated Press did was harass regular people like me and other farmers, whose only sin was to be forward thinking. So, during their quest to find some dirt within the TAFA program, the AP asked the Department of Agriculture for the files of all the companies who had defaulted with TAFA guarantees. Next, they asked for information on current participants in the program. Of course, many of the ventures were unusual, as that was the whole purpose of the program, and a shrimp farm in the West Texas desert sounded perfect. They figured this had to be some kind of scam, and of course Rick Perry was profiting from this somehow.

The problem was that the records of the defaulted loans are indeed open records available to whoever wishes to see them, but current, good loans are not subject to examination by anyone but the loan recipient and his or her bank. That is federal law: banks can't give out your information and they can be held accountable for breaches. When the Associated Press asked the Department of Agriculture for all these records, they apparently did so with some attitude. It scared the crap out of the bureaucrats at that department, so instead of deferring to the Texas Attorney General's Office for guidance, they just handed them everything they had, even our active files on a loan that was in good standing and current.

Open records requests are funneled through the Texas Attorney General's Office so they can vet and sort out the appropriate information to fulfill the request. The folks at the Department of Agriculture failed to do this. The actions of those who bypassed the process were not only illegal, but our bank had de facto broken the law because the AP, who was unauthorized, had active and current bank files they were not supposed to have. The folks at the AP were grinning like outhouse rats, and my banker was freaking out. Not only was all this illegal, but if any other bank customers found out that their information could be obtained so easily, they would have made a run on that bank as well as possibly sued them. The only choice the bank had was to file suit against the AP and put an injunction on this information to try to stop the AP from publishing anything. An Austin judge agreed that the whole thing was clearly illegal and brought the hammer down. This caused the AP and every other news agency in Texas to go berserk, noting "prior restraint" and "remember the Pentagon papers."

Now keep in mind that this was just a shrimp farm loan, and not a real large one at that. It was overreach, but with the bank filing suit to stop them, the AP assumed that graft and corruption were involved, or why else would anyone try to hide this information? The obvious reason was that no one is supposed to be able to just obtain the personal information of regular people who are just doing their thing! That file contained all my personal information and that of my whole family, as well as the folks, mostly family, who had invested in the Permian Sea Shrimp Company. They obtained birth records, Social Security numbers, and a lot of other private information. It was everything nightmares are made of when the wrong folks get your information. On top of that, they got the information illegally. That didn't matter to the AP. The only thing that mattered to them was that someone in a small town in America had the audacity to challenge them. They were upset, and now I was the target. For me it was just baffling and ridiculous.

The worst thing was that I had a shrimp farm to run and no time to deal with this. It was really between the bank and the Associated Press anyway, although that certainly didn't stop them from writing these dark and sordid stories about what a bad guy I was for taking advantage of the taxpayers, with Rick Perry as my accomplice. At that point, the Texas press writers never once interviewed me or called or anything. No time to gather facts; we have a crisis to take care of here. The fate of the state of Texas lies in the balance! While the court injunction stood in effect, the Permian Sea Shrimp Company and Bart Reid were disparaged in every newspaper in Texas. Fortunately, a quick trial date was set and off to Austin we went to fight the press in court. Just before I left for Austin, I got a call from someone with the AP. He asked me why I didn't just drop the suit and let them do their research on the TAFA. I told him, first, it wasn't my suit, it was the bank's obligation to sue; second, they didn't need my information and should not have gotten it. He told me that the information was personal and boring and nothing they could really use, but they had to win this now. He said, "No one tells the press they can't publish whatever they want to." We went to trial.

The bank's lawyer did not have much hope to beat the AP because the precedent for prior restraint was firm. They can publish whatever they have even if it was obtained illegally, which was the case here. He had hoped that maybe through this we could make a case for suing the Department of Agriculture, since they were really the culprits who broke the law by wrongly giving the files to the AP to begin with. Just before we went into the courtroom to start the actual trial, the lawyer for the AP called us into a room for a brief chat. He basically said that we should just give this up as the AP could destroy our reputations if we pursued this thing. He said they could publish stories that we all were suspected pedophiles or had mental disorders or all sorts of terrible things. We should just give in and go home, or we might be ruined forever. That's our American free press looking out for all of us. Well, we didn't go home. We didn't win either. The judge let us call to the stand the people at the Department of Agriculture who gave away the files, and she let us talk a lot. I gave an impassioned speech while on the stand about how this was not supposed to be how America worked and that people didn't go to war and die so the AP could get your personal bank records. I was satisfied I had made a great impression on all who heard it and was sure it would inspire a whole new generation of patriots.

The judge found for the AP after pondering the whole thing for maybe sixty seconds. Maybe she didn't want to be called a pedophile. The bank left very nervous about the fallout back home with their customers, but nothing ever came of it. I was disappointed that they could now use all my records, but I was also pretty sure there was nothing in my personal records that they would find interesting or useful for mudslinging. I was just glad I wasn't running for office. Ironically, soon after the trial the Attorney General released its opinion that stated in part:

> In our opinion, all financial information relating to an individual—including tax records, social security numbers, sources of income, salary, assets, medical and utility bills and credit history—ordinarily satisfies the first requirement of common law privacy . . . and should not be disclosed. In addition, we conclude

that the release of detailed information revealing Permian Sea's
financial information would cause substantial competitive injury
to Permian Sea. . . . We believe that releasing the document titled
"Loan Guarantee Program Summary" is sufficient to satisfy the
legitimate public interest.

What do you know, we were right all along, and $27,000 poorer
(since I had to pay the lawyer fees) thanks to the lawyers who win
either way. I did have a few state representatives contact me later
and volunteer to sponsor legislation allowing Permian Sea Shrimp
Company to sue the Department of Agriculture if I wanted to do
that. I was ready for it to be over and done. I was a shrimp farmer,
not a litigator. After it was all over, life went back to normal in West
Texas and Rick Perry won the election anyway. What a waste of time
and money and nerves. And the saddest part is that no one will ever
remember my amazing and heart-wrenching courtroom soliloquy.

I was happy to be out of Austin and back in Pecos County looking after shrimp again. By 2001, shrimp farms in West Texas were
producing close to 800,000 pounds per year of shrimp combined.
The Permian Sea Shrimp Company was making a real impression
on the shrimp industry. We produced the best shrimp money could
buy, and our shrimp were large and medium tails which are the
most popular sizes for the restaurant market. We had good local
retail sales and good buyers from on the coast, and we got good
prices for our shrimp. Lots of folks would come on advertised harvest days and buy fresh shrimp to take home and freeze. Some of
these folks would come into the ponds and help us pick the last of
the shrimp just for the fun of it or help us load or whatever. It was
an experience that I think folks knew they could not get anywhere
else, so they wanted to participate in it to the fullest.

That was also the year that some small-time, two-bit oil company
from Kentucky came to Pecos County and started drilling on the

Permian Sea Shrimp Company land. In some scenarios that might turn out to be a good thing, but since we didn't own the mineral rights, only the surface, it was nothing short of a pain in the butt. Now I always got along with the oil field folks around here, but these guys were something of a different animal. They did not seem to care about the folks who regulate the activities of the oil field, the Texas Railroad Commission, or its rules. They did not even extend the normal common courtesy usually given to surface owners and local people affected by oil field operations. These guys were jerks. These fellows staked out two spots on the farm where they were going to drill wells, and these spots were in an important area of our operation where we drain water. Their ultimate plan was to drill several really shallow oil wells all over the farm.

These guys did not have much money and it looked like they must have built all their drilling equipment themselves. My son's Tonka toys were better equipped to work in the oil fields of West Texas than the stuff these guys showed up with. A bunch of shallow wells was all they were going to be capable of drilling at best. I hated the idea they would be crawling over my farm doing whatever they wanted, but unfortunately that's the way the law works. The surface owner has no rights at all and has to just stand by and take it. The law says that the oil company can do whatever necessary to extract oil from the property. The law also says they can even use the surface owner's equipment to accomplish their work!

How did I learn all this, you might ask? I went to Odessa to talk to one of the craftiest and most successful lawyers that ever lived in my opinion, and he just happened to have grown up in Imperial. His name was Warren Heagy, and he was a badass. I had studied his reputation closely and knew this was the guy. In the recent past, he had a client that hired him to sue the insurance company of a local electrician friend of mine and the client won hundreds of thousands of dollars. My friend said he would for sure hire Heagy next time he ever needed a lawyer. In another case he represented a guy who was drinking in a work truck and had a wreck that killed a man. Heagy not only got a not guilty decision for the fellow, but the dead guy's insurance had to pay for repairs to his client's truck! There were many cases like these. He was talented for sure, and since he was from Imperial, I thought he might give me some sage, bulldog lawyer advice. Maybe even pro bono. I just knew he would sign on to run these oil guys out of town. Instead, he told me I better make friends and try to get along. He explained that the oil field laws were not in my favor and as good as he was, he could not penetrate the armor of this beast. It was a letdown, but I respected his advice and returned to the farm schooled and humbled.

I made a plan to go see the main owner of this Kentucky oil company, in an effort to try to be cooperative as Heagy had suggested. This guy was a short, filthy little man who always seemed angry when he wasn't bragging about how awesome he was. I informed him that I would be harvesting in a few months and the harvest water from the ponds was diverted directly to the area where he was setting up his drilling rig and other assorted junk trucks and equipment. I was quite certain that all that pond water would inundate his equipment. If it didn't sink it in the mud up to the axles, the equipment certainly would be inaccessible and stuck for a very long time. I disliked that they were there but felt he ought to understand what was going to happen with all his equipment. Even though I had no leverage to make them leave, I thought it was the right thing to do to give him a heads-up about the coming flood. I also suggested that he use his time and money to do some earthwork to divert the future flood away from his equipment or risk a perilous outcome. I thought I was doing him a favor by warning him about all this, but in all his brilliance, he told me to mind my own business. He told me that nothing I did on my little farm could affect his operations. He said he just knew there was big oil under that land, and nothing would stop him from getting to it. So, I left him to his own devices and never gave it another thought.

In the next couple of months, he managed to get some drilling done, all the while breaking so many Railroad Commission rules that the local oil field guys got concerned about what he was doing to the land and the area that they turned him in for various violations. This must have gotten him really angry—I heard he was fined $50,000, but I don't know that for sure. One day in September, just before we were going to start our first harvest, I saw him out there around a location and went over just to remind him about the water that would soon be coming his way. As I approached, he jumped off the rig with a 36-inch pipe wrench in his hand and came at me ready to fight. Now I had not been in a fight in years, but I very much disliked this asshole, so I was actually surprised a little that I was sort of looking forward to it. As he got within about ten feet from me, I said to him, "You better knock me out with the first swing of that wrench 'cause if you don't I will for sure beat you to death with it. There is nothing I want more right now in the whole world than to kick your sorry ass." He stopped in his tracks and started yelling something about my turning him in to the Railroad Commission and other such nonsense. I angrily asked him to read his paperwork from the Railroad Commission and show me where my name was listed. The Railroad Commission always lists the complainant on its paperwork (a fact that I am sure inhibits many folks from complaining). He admitted that I was not mentioned anywhere but he was sure I had something to do with it. I told him I was there to remind him that in about seventy-two hours his drilling operation would be sitting in a sea of pond water two feet deep and that he had only himself to blame as I had warned him repeatedly and he had done nothing to mitigate it. He just turned and walked away with his big wrench in hand. Seventy-two hours later, the first of what ended up being more than 300-acre feet of water during the next six weeks went washing through that pasture. They could not operate or even get to their equipment until the next spring. It was the best harvest season ever.

CHAPTER 9

THE SMALL SHRIMP FARMS

By this time there were the two big farms in Imperial: the Permian Sea Shrimp Company and the old Regal Farms which had changed its name to R&G, or Lighthouse Seafood or something like that. As I mentioned before, it was owned by my friend and sometimes-shrimp buyer Dennis and his partner. There were also three other small operators that had come into fruition in the area during the last couple of years.

TUCKER FARMS

The first little operation was just over the Pecos River in Ward County. This farm was owned by a fellow named David Tucker, who had a master's degree in chemistry and had been working in the oil field supply industries since he moved to West Texas. I had known him for a while and our sons were the same age and hung out with each other. He was always interested in the shrimp farming and had been coming around and talking about it way back since we started Triton Aquaculture. One day he came to me and said he was fed up with the oil field and wanted to try his hand at shrimp farming and asked if I would help him out. I thought about it for a while before giving him an answer. I was thinking that this business is not for everyone, and he really needed to know what he was getting into. So many people only see the fun parts, like harvest day, and they believe it's going to be a nonstop party. They never see the day-to-day farm drudgery in 112° summer heat, never getting a vacation while your kids are out of school, having to stay up all night to monitor oxygen levels and run emergency paddlewheel aerators, then on two hours' sleep start it all again in the morning. I could tell he was really stressed by his oil field job, and he was a very smart and hardworking guy, so I relented and agreed to help him start a little farm.

I convinced him to hire the Mennonites to build him four ponds about 3.75 acres each. He also hired, on his own accord, old Jeter to drill him a cable tool rig water well, and in short order he had a nice little farm. I told him that for three cents per pound of produced shrimp, I would let him use my cool shrimp harvesting pump and that would save him some start-up money. I also found him a shrimp buyer for his crop at harvest time and arranged all the harvesting details. All he would need to do is stock and take care of his shrimp all summer. He had a canoe and decided that since he only had four ponds, he would feed each of them by hand out of that little boat. This turned out to be a great idea as he was able to spread feed all over the ponds and all the shrimp got a full belly. This also made for a really good survival rate of the shrimp. His shrimp were ready to harvest earlier than the other farms' his first year and I think a lot of it had to do with feeding out of that canoe. Since his shrimp were ready by the first of September, I took my harvest pump eleven miles down the road to his farm to harvest for him as we had agreed. He had a banner crop, and since it was early in the season, he also got a good price as the market was not yet saturated like it would soon be when coastal farms, West Texas farms, and shrimp boats all started to harvest and sell.

About halfway into harvesting his four ponds, we got hit by an extremely early cold front. It killed little swallows and purple martins and other birds that were not ready for the cold front so early and had not set out for parts farther south. The unexpected cold threatened to kill many shrimp as well as it dropped the water temperatures very low, very fast. I got with David, and we decided that we needed to just hit it and do a harvest marathon for the last two ponds so that I could get my harvest pump back to my farm and start harvesting before the weather got any colder. I was afraid I could start losing shrimp if it turned out to be an early winter. We arranged to have a truck come ASAP with plenty of totes and enough ice to harvest the last two ponds. We got busy and in two and a half days we got those two ponds harvested and in the ice. By the time we finished we were absolutely beat. On that last pond, our good friends the Mennonites came to watch the harvest, which was winding down at just before sunset. Our friends had a bunch of relatives and other Mennonites visiting them, mostly young married folks. I think there were about ten or twelve women and maybe eight or ten men. We were worn out, but it was still quite an event and part of what made harvest time so enjoyable. After a summer of working night and day it was nice to be around other folks and celebrate the spoils of our labor.

I got my shrimp pump back to my farm and proceeded to harvest without any losses from the cold, but my banker was concerned. He was worried about the risk and told me that in the future I had to harvest my own farm first and once that was completed then I could help someone else or rent the pump or whatever, but only after I was finished on my farm. I had to tell David that the best plan moving forward was to build his own harvest pump like the one I had built. He would just have to be on his own from then on. Of course, this also ended the three cents a pound deal for helping and finding the buyers, but I didn't really have much of a choice in the whole thing.

ARNY'S FARM

That same year that David got his first crop in, another local fellow made an attempt at farming shrimp at the old water district farm. That farm had sat idle for many years, and it was a bad idea at this stage to try to make a crop there, but he did try anyway. This local guy, Arny, fancied himself a farmer, and for the whole time I had lived in Imperial he had tried farming this or that, here and there. I always saw him plowing a field or planting seeds or working on farm equipment. When the shrimp farming came along, he just had to try his hand at it. I think he plotted and planned for quite a while and then finally got the nerve to give it a try. That year he leased the old water district project, filled up the ponds, and stocked some baby shrimp. I never really talked to him much that summer because I was so busy with Permian Sea Shrimp Company and David's farm. I had no idea what kind of survival he was getting or how big his shrimp were. What I did know is that the same early cold snap that came in while we were harvesting David's farm killed everything on his farm.

I never learned how many shrimp he actually had, but whatever he had died from cold water temperatures in his ponds. I believe those shallower ponds on that old farm just got too cold too fast. I also doubt he knew much about how to manage the warm well water to keep the shrimp alive. The well water is about 72° Fahrenheit year-round, and while the wells can't pump a lot of water compared to an entire four-acre pond, you can run the water wells in ways that can keep your shrimp alive until the cold passes and the days warm back up a bit. It took me years to learn how to do that, and I am sure his naiveté contributed to the loss of all his shrimp. I tried earlier to convince him not to get into this business, but he would not listen to me or anyone else. I figured, hey, it's a free country. The shrimp loss and the financial loss were sad for sure, but things can and do sometimes work out for the good. The poor dude gave up farming of any kind and ran for justice of the peace and won. He had a good, stable county job without the risk of Mother Nature. He never had to farm again.

D&T FARMS

The other little farm that started up a couple of years after David started his farm when Permian Sea Shrimp Company was about four years old was a farm called D&T Farms. This farm was started by a Mennonite fellow and his wife who found their way here from Kansas. The Mennonites who built our farms had been in Imperial for several years and wanted other Mennonites to come in so they could socialize and have a church and school for their kids. They don't do well alone in the desert, even among those of us whom they enjoyed being around. They need other Mennonite friends to encourage them in their faith. I supported that completely and wanted to help them bring in others. I showed my farm and the local area to many visiting Mennonites who wanted to check it out and see if they might consider life in the desert.

I remember a group from somewhere east, maybe Pennsylvania, that came on such a visit. Seems the older folks in a Mennonite church or community often encourage the young folks to move on. Maybe they do this for missionary purposes or to improve their biodiversity, if you will, which can be an issue in closed societies. For whatever reason these young folks were urged to come to West Texas, and they certainly did not seem enthusiastic. Upon further inquiry, I discovered that the men in the group had heard that West Texas was Comanche country, and they had all brought firearms to protect the womenfolk from the savages. The ladies were not pleased that the men had to make such expensive purchases just to go on a tour of the country and the men were equally not pleased that they may have to resort to arms just to survive this forced excursion. Neither side was at all excited about moving to Imperial, and I am sure they spent whatever downtime they had studying the map for the quickest way back home.

We also toured around some Amish relatives of our Mennonite friends. That was an interesting and enlightening visit. I learned a lot about Amish folks that I never knew before. For instance, that there are two major and very important divisions of Amish—those that use buttons on their clothing, and those that believe buttons are sinful (don't ask me why); therefore, this group only uses pins to hold their clothing shut. When my friends told me about this division in the Amish faith, I really thought they were pulling my leg. I looked all throughout the Bible and never found one reference to buttons but then again, those robes and tunics we assume they all wore back then didn't have much call for buttons. I guess their interpretation is that if Jesus didn't use buttons, then they are off limits. Jesus did not eat pizza either, but I am pretty sure he would be cool with pizza, and buttons.

Our Mennonite friends often joked about their Amish relatives and frequently had sarcastic things to say about how they lived. My Mennonite friends did not seem to take very seriously the Amish practice of self-denial and primitive living. I once mentioned that I thought it might be difficult for Amish folks to live without life's conveniences. They immediately dismissed my concern and adamantly told me that to the Amish, life was a big, long, fun camping trip, and we were the suckers for thinking that somehow they were all uncomfortable and suffering in their chosen lifestyle. Anyway, to my astonishment, the button vs. pin doctrine dilemma was real. It turns out that my friends' relatives from the Amish faith were from team pins.

I also learned that while Amish will not own a motorized vehicle, they certainly will ride in one. Apparently large numbers of Amish are keen to take long vacations throughout the country every year and a very lucrative industry exists of drivers who are paid to take the

Amish all over the place in vans and buses. Such was the case when an eighteen-seat passenger van pulled into my drive full of Amish dressed in their Sunday best. Of course, they wear similar clothing every day, so I can't be sure if it was their Sunday best, but they looked nice and pressed so I just assumed it was their Sunday best. Yep, these were the relatives from back East, and just as described, their clothing was fitted with pins in all the places that a button, snap, hook fastener, or zipper would normally be found. The pins were shiny and new looking. I could not stop staring at them. Pins holding shirts together, suspenders attached, pants closed and keeping dresses from falling open. The women wore dresses that were even more plain, if that is possible, than the ones our Mennonite friends wore. The men, however, had stunning maroon-colored, silk-looking shirts that in my opinion were anything but modest and simple. The men all had big leather boots and fine black wool hats with black hat bands. It was quite a sight for this part of the world, and it all looked very hot for West Texas in the summertime.

My Mennonite friends showed up to reunite with their relatives, and we all toured the farm. I was super interested in seeing those pins work in action. It seemed to me a risky way to keep your clothes on. While we walked around the shrimp ponds and as I gave my nickel tour of the place, I kept one eye open for the wardrobe malfunction that I was certain was going to happen. I mean, they were held together just by *pins*! And, if there was a wardrobe malfunction, what was the protocol? I didn't really want to see, did I? Don't judge me, I'm a scientist. I'm just curious.

Of all their relatives, friends, and interested church people that came to visit our friends, only one other Mennonite couple ever decided to move to Imperial, and it was this couple, Dale and Tina, that started D&T Farms and who gave up their Kansas corn farm for the West Texas shrimp farming life. They built a little four-pond farm almost identical to the Tucker farm. They would use David's shrimp pump for harvest and feed from a boat just like David was doing. Better yet, they had two or three young sons to help with all this so labor was plentiful and the future was bright. Their first year they produced an extraordinary number of small shrimp. Since these guys were not trained in aquaculture at all I assumed that they just did not know how many shrimp to stock and so they ended up overstocking their ponds. The shrimp had good survival, and Dale and his boys did a good job their first summer, but the ponds just had too many shrimp. And worse, the bumper crop of shrimp suffered what we call the aquarium effect. You know how you can take a little goldfish and put him in a goldfish bowl and he stays a little minnow, but if you take him out of the goldfish bowl and put him in a pond, he then becomes a big ole ugly carp. That is the aquarium effect. The animals grow to fit the space they have and the available food supply. If you put in way too many shrimp and they survive, you will get many small shrimp.

The problem is that small shrimp are not particularly valuable, and D&T Farms had to work very hard the rest of that year to sell their shrimp. The buyers would not take small shrimp, so they had to have them processed and sell them on their own, all over West Texas. It was a tough sell. They hatched up a story that small shrimp were more tender and tasted better to try to convince the public to buy their shrimp. We all laughed since we knew in time they would produce large shrimp and then would have a tough time undoing their own myth. Sure enough, the next two years D&T produced nice, normal, large-size shrimp and had to do a lot of backpedaling with the customers they had convinced only a year previous that small shrimp were better. The large shrimp sold well; I don't think anyone ever really thought small shrimp were better.

D&T Farms had about three good years, but they ended up closing down and moving away. I believe it was as much for personal reasons as financial. Other Mennonites never came to the area, and they were very lonesome for other folks of their faith. Moreover, there are much easier ways to make a living than shrimp farming, so they handed the keys to the farm over to the bank, headed down the road, and never looked back. Our other Mennonite friends that had lived in Imperial so long and had done all the dirt work for all the farms also left West Texas. They had lived away from others of their faith long enough and they had young children that needed to be schooled in a Mennonite school and raised amongst those of similar beliefs. It was very sad for us to see our friends move away after so many years in the area, and we certainly understood why, even though it was painful to see them go. They had been integral to the formation of this industry, and it would not ever seem the same without them.

In all, the small-time farms existed for about five years during the height of the shrimp production in the area. The big farms produced far more pounds of shrimp, but the little farms had some good luck and contributed to the overall production in the area. At its peak in about 2001, the shrimp industry in West Texas produced more than 800,000 pounds of shrimp per year, and we were all pressing to get to a million pounds per year. We were a major force in the seafood industry and now had respect and some clout. Some of the farms felt that if we were able to form some sort of a shrimp farm cooperative we could get better pricing on feed and equipment. We might also be able to put all our shrimp together and get better prices from the buyers on the coast. This is the main reason farmer cooperatives exist in the first place, to have a legal way to control prices. A legal monopoly. We put together a cooperative called the West Texas Desert Shrimp Farmers Association and started the legal work to make it official. The other big farm, R&B, was not interested in being a part of it since they owned their own processing facility and distribution back down on the coast. Without all the farms in, the deal just wouldn't work. We went to some seafood trade shows and put the name of the cooperative out there, but we didn't get the attention of the industry nor did any of us have time to organize and run the thing like we should. We all had our shrimp farms to take care of and that alone is a lot of work without taking on something of that magnitude. What I always thought we should really look into was our own processing plant. This would accomplish the same things as the cooperative but with even more control of our product. The problem was that a processing plant takes a huge amount of capital to build from scratch, and none of us had the cash to spare or the ability to borrow any more than we already had. It was just going to have to be business as usual for a while, but that was fine. Prices were good and we could sell everything we produced. The dream had become reality and life was good in West Texas for dreamers and for shrimp farmers.

CHANGE AND SURVIVAL

It is true that the one thing that is for sure is change. What I had envisioned those years before had finally come to fruition and we were making good money, raising our families, and having a pretty good run. Then everything changed, and none of us saw it coming. The attack on the World Trade Center and the Pentagon on September 11, 2001, now known as 9/11, was a horrible event in America and all the innocent lives lost were certainly a much greater tragedy than anything we were forced to endure so far away from the actual event. There were, however, big repercussions for the shrimp industry caused by 9/11, and the effects are still with us even today some twenty-plus years later.

In the 1980s and 1990s the shrimp business flourished in America. The wild-caught industry on the Gulf Coast was huge, and there were enormous shrimp fleets out of Aransas Pass and Brownsville, Texas, as well as hundreds of boats in Florida and other gulf states. The market had gotten very strong, and demand was enough that it was finally possible for the shrimp farming industry in the United States to compete with the wild-caught fleet. Imports of shrimp from other countries were growing but were not significant enough to affect the industry at that time. In the hours after the 9/11 attack, all the ports in America were ordered to close. Nothing could come in and nothing could leave. There were, in the ports of America, ships with container loads of imported shrimp from all over the world, but they were mostly from China. Chinese shrimp was considered inferior and was not commonly sourced by any fancy restaurant chains or by the retail grocery chains. It was mostly imported for food service and fast-food chains. After 9/11, there was much uncertainty in the country about what the government would do with all the cargo and ships in the ports. Cargo might be held up for months, risking spoilage and ruin, or it might be turned away and forced back to its port of origin. At the first chance they got, the shrimp import companies offloaded their cargo and put it on the market at rummage sale prices. They were not looking to make money; they were just in a panic to get that product off the ships and into the market to bring in whatever they could to offset expenses and avoid any further costs.

All of a sudden, the shrimp market was inundated with cheap, mostly Chinese, shrimp. The buyers took to it like junkies to free crack—and like a bunch of junkies, they got addicted to it as well. Soon the folks who in the past would never be caught with imported shrimp in their restaurants were snapping it up by the container load and totally ignoring the domestic shrimp. American-farmed and wild-caught shrimp both were left in the cold. Because of their government subsidies, China could grow, process, pack, and ship shrimp at prices that wouldn't even cover our shrimp feed prices in America. Almost overnight, we couldn't give away shrimp. What really made this especially bad was it all happened right at the beginning of our two-month harvest season. I was literally preparing for the first pond of the season the morning those planes crashed into those buildings. All shrimp prices crashed too, but now West Texas shrimp became almost worthless. No processing plant or broker would bother with the expense of sending a truck with ice all the way out to Imperial when they had cheap foreign shrimp just a phone call away.

We had to think fast to survive. The other large shrimp farm in West Texas and at least two of the small farms would not survive it. Many shrimp farms on the coast and most shrimp fishing companies were all gone, out of business or bankrupt, within the next eighteen to twenty months. I had heard that banks started to refuse to even take back shrimp boats anymore because they had repossessed so many and couldn't get rid of the ones they had. Congress passed a law allowing for tariffs to be put on certain foreign shrimp due to dumping and with something called the Byrd Amendment in place, the tariff money should go to the affected producers like shrimp boats and shrimp farms. The Byrd Amendment was one of the few things Congress has ever done right. This law provided that if tariffs were imposed on countries for illegally hurting industries by dumping or price manipulation, the money from any imposed tariffs would go to the industries that had been negatively affected and not into the government coffers. But as usually happens with government programs, the wrong people got the money. It went mostly to importers and processing plants that were importing the foreign shrimp in the first place! Almost no farms or boats got any tariff money. Then to add insult to injury, Congress later repealed the Byrd Amendment. With the repealing of the Byrd Amendment, the money from the tariffs would go to the general treasury and nothing to the industry affected. The benevolent politicians decided that negatively affected businesses and the people who work in them could just apply for scholarships and assistance to learn another trade. Thanks, government!

Due to the situation the industry was in, we decided we would risk it and pay for the trucking to the coast and processing fees ourselves. We would then retain the finished product and try to sell it however and wherever we could. This was a big undertaking. Selling three or four thousand pounds on the pond levee at harvest time was one thing, but moving 300,000 pounds or more was a whole other ballgame. My plan was to start wholesale shrimp sales to restaurants all over West Texas. Everyone had heard of the Permian Sea Shrimp Company and West Texas shrimp by now and it seemed that this would be a way to stay afloat. It meant a lot more work, and we would need a place to store all that frozen shrimp while we sold it throughout the area. There had been an old convenience store in Imperial that had closed years before and had sat vacant ever since. A bank in Fort Stockton had foreclosed on that store long ago, and probably after all these years would be ready to make a deal with someone to take it off their hands. I took a trip to town and in less than two weeks I had gotten a loan to purchase and renovate that store. This store had a very large walk-in cooler on one end that could hold thousands of pounds of frozen shrimp. I figured we could convert that big cooler to a freezer and store the shrimp there. It would make a good location to work out of while wholesaling shrimp all over the area. It was a quick act to try to pivot and keep the West Texas shrimp industry relevant and extant. We had to get rolling, to keep the industry alive—or at least a farm or two. It turned out that it was actually an easy sell, and it worked very well. We picked up restaurants and retail stores from Andrews and Midland down to the restaurant up in the Basin at Big Bend National Park, several in the Terlingua and Lajitas area, and southeast all the way to Del Rio. We had our shrimp processed on the coast and then trucked back to Imperial, keeping the freezer in the old convenience store brimming full of shrimp inventory. We delivered north on Tuesdays and south on Thursdays.

I didn't think we would ever use the main part of the old building; it was just the freezer space I needed—or so I thought. My wife had the idea to serve shrimp plates for lunch on the days we were not traveling around West Texas delivering shrimp to stores and restaurants. As if running a shrimp farm and a shrimp wholesale business weren't enough. But then we were young and hungry.

CHAPTER 10

THE SHRIMP STORE

t seemed like a simple enough idea and, who knew, it might generate a little extra money. Boy, that was an underestimation! People went crazy for it, and we could not keep up regardless of how hard we tried. They came from everywhere wanting shrimp to eat and shrimp to take home. Patsy and I discussed the whole situation and decided that if people were that enthusiastic about a shrimp restaurant, then we had to go for it and give them one. The Shrimp Store was born.

The idea that serving shrimp plates in Imperial, Texas, would turn into some big, popular thing was as far from our minds as possible. Imperial is not on the edge of the world, but you can damn sure see the edge from here. We blew those three all-important rules for the restaurant business—location, location, location—completely out of the water.

We did know that to have any sort of real restaurant we had to make this place look presentable and be able to pass the county health inspection that all eateries in the county had to pass. The old building sat vacant for almost ten years prior to our purchasing it and I had not planned for any serious renovation of anything but the coolers. We went to work on the old place. First, we needed a sign, which I made from sheet metal and angle iron. On that,

I placed some big vinyl letters I had purchased reading simply, "Shrimp Store." The store had to be obvious from all directions, so I had an idea. I would take two old water well pipes, screw them together, and then put the sign on the very top. These pipes were ten feet long each and ten inches in diameter so they would make a sturdy and magnificent sign pole. We screwed them together and erected them in the parking lot then realized we had no way to get the twenty or so feet to the very top and mount our sign. Luckily, our water well service guy, Randy Hartman, happened to be in town with his water well service truck, and he and I hatched a plan. We would back his truck up to the mounted sign pole and raise the derrick up. Then we would put my wife Patsy in a harness and with one cable, called the sand line, lift her up to a position to mount the sign. Then we would use another cable on the derrick to raise the actual sign up to where she could screw it onto the top of the well pipe. As you can imagine Patsy wasn't all too thrilled, but she took one for the team. The plan worked like a charm. Every human being in the area stopped to watch her dangling in the air screwing that big ole sign onto the threads of the water well pipe while twenty feet in the air. Nothing like making a spectacle to get folks' attention, and attention was what we were going to need. Now we stood out, and people were curious about what we were up to.

As for the inside of the store, we cleaned it up and waxed the floor. We bought a few plastic tables and folding chairs from Sam's Club and the dining room was set. Fortunately, the old store had had a deli and kitchen area, so rigging it all up to cook was simple enough. We just needed a vent hood and some HVAC refurbishing on the deli fridge, and we were good to go. We didn't buy a bunch of fancy cooking equipment because we only planned to serve a handful of plates at lunchtime. And we thought the lunch plates might also lead to some retail sales of frozen shrimp.

By week three we were serving more than seventy-five people a day at lunch and had to hire an employee just to keep up. We realized that this thing had potential, so we decided that we would take this seriously and make a go of it. We changed the hours to serve lunch and supper all week long. I would deliver shrimp during the week (and run the shrimp farm) and work at the store when I wasn't doing that. Patsy signed on to work full time at the store and hired two more employees to help her run it. We had seen many ventures be successful and many ventures fail, and we felt we had a good handle on why some restaurants worked and others didn't. I hoped our observations were correct because we were already starting out behind the eight ball in probably the worst location in history for a restaurant. We needed to get this right.

The most important two things that we felt were critical to success were novelty and simplicity. We had the novelty down—a shrimp restaurant serving homegrown shrimp in the West Texas desert. Pretty novel. Now simplicity. We had three main dishes— boiled shrimp and sautéed shrimp every day, and fried shrimp on Fridays and Saturdays. We also served hamburgers with hamburger meat from an organic beef farm down at Fort Davis, and a local guy smoked us brisket to make burritos for the non-shrimp eaters. My wife made amazing coleslaw and potato salad that were the main requested side dishes and she made awesome, homemade desserts that were as famous as our shrimp dishes. We would also do a Wednesday special each week to change things up a bit. This would be a specialty shrimp dish like Shrimp Imperial: poblano chiles stuffed with shrimp and cheese and wrapped in bacon and grilled over a mesquite fire. In addition, our shrimp burrito was addictive and very popular.

The whole thing was a phenomenon. The parking lot would be full and there would be cars across the street along the irrigation ditch. We even had oil company CEOs in helicopters land near the school where there were some open fields and walk over and eat with us. The weekends were crazy busy, and by closing time on Saturday night we were beat. No one would have ever predicted this would happen by any stretch of the imagination, but I think it was a testament to how great the shrimp grown out here tasted and to my wife's amazing dedication and skill in running the place. I think the people came for the shrimp but left feeling they were a part of the whole thing. That may sound corny, but perhaps that is only if you never met Patsy.

Even the locals started coming in every day. Oil field pumpers, mechanics, ranchers, and more pumpers. In the beginning, they were there mainly for the fact that the metal building in which the Shrimp Store resided was like a super shield against cell phone signals. They could hide in the restaurant for hours without their cell phones ever ringing. No bosses to check up on them, no wives to give them lists of things they had to go thirty miles out of their way to pick up before heading home. This, combined with the good food, became their nirvana. It did not take long before they were comfortable enough lounging at the tables that they started in on rumors and gossip of all sorts. I never saw folks that knew so little about anything important but damn sure knew everything about other folks' lives. Once each member had professed all they knew about the private lives of everyone they knew, they would move on to grander subjects, giving great orations on politics and religion, or their distaste for both. They discussed the weather as if they were meteorologists and current events as if they had witnessed the occurrences. They would sit for hours after the lunch rush deep in discussion. I called them "The Imperial Brain Trust and Mensa Society—where the world's biggest problems are solved by the world's smallest minds!" At least they paid for their meals and never got so loud as to run off the other customers. They were harmless, and their trucks parked outside for hours made the place look busy, which is always a good thing for a restaurant. People pull into cafes where there are loads of vehicles and it looks like that's the place where all the locals congregate. Special of the day: shrimp and gibberish.

A DESERT DESTINATION

Business increased every year that we were open. Our advertisement was mostly by word of mouth because we couldn't really afford any real advertising, and besides, where would we advertise? Everyone in West Texas knew us.

One day out of nowhere we got a visit from a famed restaurant critic named Dotty Griffith who wrote for the *Dallas Morning News*. She did not warn us of her visit or anything, just showed up one day to see what the fuss was about. We didn't even know who she was or that she was going to write a review about us until she was ready to leave. I guess that's how she wanted it. Raw, unfiltered and no bullshit. She wrote the most glowing and wonderful review of our food and service and that article just set everything on fire again. We started to get visitors from all over Texas after that review. We were very grateful to her for it. After her visit we got a little *Texas Monthly* magazine write-up. These things were important back then. This was before Facebook and Twitter, when people actually read magazines and newspapers and watched shows on television. That was where we all got our information. Then one day the big one came!

We got a call from Bob Phillips with the television show the *Texas Country Reporter* to see if he could do a show on the farm and the store. This was a big, big deal in Texas. The *Texas Country Reporter* was enormously popular. Being featured on that show made you a part of Texas folklore, and there was no greater Texas folklore than farming shrimp in the saltwater desert of West Texas and running the best shrimp eatery in the same location. You have to be some kind of original for Bob Phillips to do a story on you, and you have to be something really Texan too. I was flattered and frankly beside myself. I had been a fan of that show forever and I also knew that it would really put us out there in the Texas mainstream. The crew for *Texas Country Reporter* came to the farm right in the middle of harvest season in 2002. We had beautiful weather and good ponds with lots of shrimp to film. After they filmed the harvest on the farm, they went to film the Shrimp Store. There were lots of customers who were there to support us, customers who wanted to be on TV, and a few people who were just there to see Bob Phillips. I didn't blame them; the guy is a legend and has real rock star status in Texas. I was just stoked that he did a story on our shrimp farming. Even though CNN had done a national news story back during the very early days of the Triton farm, having the *Texas Country Reporter* come out seemed more special and made us all feel very good and validated. The video of the story is still available from the *Texas Country Reporter* from 2003, episode no. 814 on DVD.

By this time, we were running the store six days a week and still delivering to many other restaurants and stores throughout the Trans-Pecos region. We also had applied for and gotten a "Sustainably made in the USA" certification from the US Park Service which allowed us to be a vendor to all the federal parks that wanted to procure shrimp. It was very good business and we shipped out thousands of pounds of shrimp each month to many of the federal parks all over the United States. West Texas shrimp could be eaten from Oregon to Maine and south to Florida. Car clubs, race clubs, and many motorcycle clubs rode to the store each weekend for lunch. The very first motorcycle club to come to the store was quite an eye-opener for us. The club president called us and told us that they wanted to drive up one Saturday afternoon and eat and hang out for a while. There would be about fifty of them, give or take, according to the guy on the phone. He told us he just wanted us to be aware and ready so that we would not be caught off guard when they showed up. This was exciting to us. To others around town this was just plain scary. I was only scared because we had not served such a large group before. We had just started this restaurant and we were a bit green when it came to serving big groups. To others around town, it was worrisome because it was a motorcycle group after all: Were they a gang like the Bandidos or Hells Angels? Were they coming to town to raise hell and wreak havoc? We were certainly newbies to all this and had no idea what to expect, but in general Patsy and I were looking forward to it.

In contrast, there was great concern by the husband of our only employee and concern by the local constable as well. After all, Imperial was a quiet town and never had much trouble from outsiders. Oh, they get plenty of trouble from the drunks and addicts that live in town, but those folks are our hometown problems. Outsiders are a whole other deal. These motorcycle gangs could be trouble, they thought.

The big day came, and just before the arrival of the would-be mischief makers, the husband of our employee showed up and demanded that he be allowed to sit behind the counter to protect his bride during the mayhem that he was sure was going to happen later in the evening. This guy was not the smartest fellow I ever met, and I am pretty sure without his wife he couldn't even dress himself in the morning, but I cooperated and let him take whatever position he thought most appropriate for the security of his spouse. Hey, I'm a nice guy, and besides, I thought, if all else fails, I could maybe use his body as a shield later if things went sour. The local lawman drove up and parked his truck at the edge of the parking lot where he could oversee all the action and respond when necessary. Thermos of coffee on one side of him and his trusty .45 locked and loaded on his opposite hip. The tension was rising as certain death and destruction was surely coming . . . but Patsy and I had a business to run, so we stayed busy making food and preparing for fifty or so people. I was certain beyond a doubt that they would eat before killing us all.

At about 6:30 p.m., the bikes started rolling in. The parking lot filled up to overflowing with magnificent Harley-Davidsons of all models, configurations, and colors. They were loud and beautiful. As the bikers started spilling into the store, we were astonished. We realized that we knew most of these folks. My eye doctor, the guy who runs the pecan orchard south of Fort Stockton, my mechanic, and lots of empty nester couples who were ready to live life again now that the kids were gone. Middle-aged, single women finding their inner badass. Older, single dudes looking to revive their long-lost inner badass. Even the county judge and his girlfriend were in the group. The county judge, for crying out loud! I got so tickled I could hardly focus enough to prepare the orders. These were the gangsters these local goofs were afraid of? Wow, I realized we really did live in a sheltered and isolated town. The night ended up as a real blast. Word got out of our hospitality, and we became a standard stop for car clubs, motorcycle gangs, RV weekenders, and even bicycle groups. We were also always the go-to location for the drivers' dinner on the night before the Big Bend Road Race, an annual road race that runs between Fort Stockton and Sanderson. We soon became a must-see destination for travelers in all of the Trans-Pecos and far West Texas area.

Another group of folks who came to see us regularly were the old people from various assisted living communities in the Odessa and Midland areas. These folks are always looking to take a road trip and see some new scenery, and I didn't blame them. I am sure life at the retirement home was incessantly boring. They would come by the busload to eat at the restaurant and often they would ask if they could tour the shrimp farm and see the ponds. I did this a few times, but after one scary incident at the ponds I had to stop giving them tours of the farm.

The incident is somewhat funny now, but it scared me a bit at the time. It appeared most of the tourists from the assisted living homes were elderly women. They always wanted to go see the farm and walk around the ponds and take in the whole experience. This was usually an uneventful tour. I mean, from time to time you would have some embarrassing situations like the time my female dog was in heat and our male dog found her among the old tourists on the pond levee. He started doing his thing right then and there. It's kind of hard to give a mini-informative lecture and keep a straight face when two dogs are humping away right near your feet.

But one incident was a little more serious than dogs with bad manners. We took a bus of elderly ladies out to the farm for the usual two-cent tour. Keep in mind, all these ladies had on lots of perfume and hair spray, and they all seemed to be wearing brightly colored sweaters and blouses. They took this trip to the shrimp farm and store seriously. We drove over from the store to the farm and parked the bus. The ladies got out of the bus and walked around the pond for a while as usual. Then I gathered them up for my brief oration on all things shrimp farming when all of a sudden, a huge, angry swarm of bees came buzzing through our location. I guess they saw all those flower prints on the clothing or smelled their perfumes, but whatever it was, the bees liked it—and they covered all those poor ladies. Of course, the women started squealing and swatting and carrying on, so the bees started stinging. Now these were not African bees or anything, so it was not like we all ran for our lives, but it was uncomfortable and painful.

We got all the bees swatted away and all the elderly women back on the bus, but I was pale with fright. I just knew someone was going to die from this and their family would sue me for everything I owned. I monitored everyone until we got back to the store, and I was surprised at how well they handled it all. I think every one of them got stung at least once but none of them seemed to have any long-lasting effects nor did their cheerful demeanor change in any way. Those West Texas ladies are tough. I was amazed. Even though none of them seemed to have too big a problem with the bee attack, I never had the nerve to take tours of old folks back out there again. The thought of what might have been just shook me too bad. I am not nearly as hardened and tough as those ladies, I guess.

Folks who came rolling into the store parking lot with big RVs were special to me. We got lots of those folks traveling through, and we liked them because they usually had lots of disposable income. Heck, those vehicles are expensive. I figure you must have some money just to own and maintain one. One time a big monster Entegra motor home pulled into the parking lot and it was pulling behind it a full-size Humvee. There was about a million dollars of rolling stock sitting there in our parking lot. Out of that motor home comes a six-foot-tall, beautiful Russian blonde who slides inside the store and asks about the fresh and frozen shrimp we had for sale. I told her the price was $5.00 a pound. You would have thought I asked for her firstborn child. She carried on about how it was so expensive and "How could we expect people to pay such exorbitant prices?" Why, we were practically thieves for asking so much for shrimp. I just smiled and pointed out the window at her rig out there, which was literally worth more than my store and my house put together. I said to her I am pretty sure you can handle $5.00 per pound, ma'am. She realized the irony of it all, purchased about fifteen pounds, and went on her way. I never understand why people who make it a point to be so obviously extravagant somehow think you're going to believe they are frugal because they complain about the price of some trivial little thing. It reminded me of the Florida doctor who always did the same thing. I remained unimpressed.

Another RV-driving fellow made our heads spin and also made us realize how popular our shrimp had become all over the country.

This guy pulled in and parked his big motor home in the parking lot. Someone noticed it had Minnesota plates and made that known to everyone. This was in July or August and this seemed strange. In Texas, we see many folks from the northern climes, but almost always in the wintertime. They come to Texas to escape the frigid temperatures up north. We call them snowbirds. To see Minnesota plates on an RV in West Texas in the hot summer was a bit strange. The fellow came in and asked to purchase twenty pounds of our frozen West Texas shrimp. We fixed him up, but before he walked out my curiosity was just too much, so I asked him what someone from Minnesota was doing in Texas this time of year. He said to me, "I came to buy your shrimp, and now I am going straight back to Minnesota! I ran out back home and needed to get some more." This guy literally drove all that way from Minnesota to Imperial, Texas, just to purchase shrimp from us. I was dumbfounded. We were all dumbfounded. We were also very flattered. It made us realize we had really gotten something special going. It is something I didn't fully appreciate the magnitude of until someone drove 1,500 miles just to buy our shrimp. Hey, when you gotta have your shrimp, you gotta have your shrimp!

To say that people came from all over is an understatement. We had a very diverse clientele and people of all stripes came to enjoy our shrimp. When you meet so many people from so many different places and so many backgrounds you learn a lot. You get to hear about and share loads of different experiences and stories with people. You get exposure to things that most people never get a chance to even know about in their entire life. It makes you feel connected in a personal way to a community of people, and it truly expands your worldview and knowledge of many diverse things. Most of the time we were the ones sharing our desert shrimp farming experiences with others, but sometimes it was we who got to learn a few new things. In one distinct case that I will never forget, I got to gain some in-depth knowledge of something I had been intensely interested in but never thought I would ever get to experience firsthand. It has little to do with shrimp farming, but it was the Shrimp Store and the novelty of the shrimp farming that made the whole situation happen.

One of my interests is hunting, especially bird hunting. I hunted quail, doves, pheasants, and whatever was in season. As someone interested in bird hunting, by default I was interested in bird dogs. At the time, I was keenly interested in duck hunting. West Texas has two great advantages for duck hunting. One is that it is located smack in the middle of the Central Flyway for migratory birds. The sky in West Texas in the winter is a literal highway for ducks and geese and other birds headed south for the winter. Just as the snow-birds from up north are driving the interstates headed south, so too are the feathered birds overhead. The other advantage is that there is almost no water in this flyway for the ducks to land on to rest and feed. When they can find water, they come into it by the hundreds.

There were only a couple of places for ducks to find water in north Pecos County: the Pecos River and the Permian Sea Shrimp Company ponds. We saved and recycled our water each year as much as we could so at any given time in the winter, we would have eight or ten ponds full of water. The ducks poured in like Parmesan cheese falling onto a pepperoni pizza. It was nirvana for a duck hunter. My son always hunted with me and by the age of fifteen was an expert on all things duck. He could call them in with wooden duck calls, he could identify what species they were from 150 yards away, he could shoot them with great accuracy, and he could eat his weight in them. I always joked that hen ducks paid my son to teach their babies how to quack. He was good at it, and duck hunting was our wintertime passion.

My son and I soaked up anything and everything about ducks, duck hunting, and duck dogs. We had a great duck dog, a Brittany

Spaniel that would hunt with us every morning, and despite not really being a duck retriever by breeding, this dog was magnificent at it anyway. It was fun to hunt with my son and that dog. One day I read an article about Black Ducks from way up north in Canada. Seems this is the main sporting duck in the Northeast and is the duck that is hunted almost exclusively in Eastern Canada and Maine. Our ducks were mostly Mallards and Redheads with some Pintails and, my favorite, the Canvasback. There was a sprinkling of other species, but these were the big four in our world. Reading about Black Ducks was fascinating to me, but what really got my attention was the part of the story of how they hunted the Black Duck. They have a secret weapon that plays on the curiosity of the Black Duck. Now this curiosity in itself was an oddity to me in that all the ducks in our area seemed to have no curiosity at all, just total fear. Even when you called them in expertly, as my son would do, they were still very skeptical of everything and would spook and fly away at the slightest movement.

Black Ducks are apparently very curious for some reason. We read that, to exploit the curiosity inherent in the Black Ducks, the Canadians had developed, through many years, a breed of dog that not only was made for hunting Black Ducks, but one that also has the coolest name for a breed ever: the Nova Scotia Duck Tolling Retriever. That is a cool name, and a cool dog. The article I read did not have a picture of the dog, but it did describe the breed as built somewhat like maybe a small Golden Retriever, but they are colored bright red with the same markings as a red fox. They have a red tail with a white tip and a white line on their nose and chest. It said they look just like a red fox but larger. A dog that has the same markings as a wild animal is interesting enough but that is not the half of it. The way they hunt is crazy. Normally, you train your dog with what is called blind manners. This is where the dog knows to sit quietly and still in the blind until the ducks fly into range. After the shot, the dog jumps out and goes and gets the ducks that were shot and brings them back to the blind. With Black Ducks you also hunt in a blind next to the water, but you let the tolling dog run all around the water's edge. Heck, the hunter might even toss a ball or a stick to him, so he plays fetch and stays active running around near the water. I would scold my dog if he acted like this!

The really odd part is that the Black Ducks find all this activity by the dog very interesting, so they either swim near to where the dog is playing or may even fly close in for a better look. This puts them within shooting range of the hunter, who then shoots the ducks, and the dog goes from playing ball to retrieving ducks. Once the dead ducks are delivered to the hunter the game starts up again. Play and shoot, play and shoot. Weird. I promise that this sort of behavior would have all the ducks we hunted long gone for sure. This was why it was so interesting to me, and I just could not get it out of my head for weeks. I studied up on the Black Ducks and learned everything I could about them and this very rare breed of hunting dog. About a month after I first read that article and sort of fell in love with the idea of this dog, I had to put it out of my mind and move on. I had plenty of other things to do running a shrimp farm and a shrimp store. The whole thing just got filed away in the ole brain under interesting but, sadly, unrealizable.

Then, one day about noon, while I was busy charcoal grilling the Wednesday special out in front of the store, a small motor home pulled into the parking lot. A super nice man with an interesting accent got out and struck up a conversation. I told him all about the shrimp farm and the store. He told me he had been traveling through the Southwest and had heard about the shrimp and wanted to stop by and see what the fuss was about. I ushered him into the store, and he sat down for a West Texas shrimp lunch. When he was finished, he thanked us and complimented the shrimp and started to head to his RV. Just trying to be friendly and before he got out of earshot, I asked him the purpose of his travels through the Southwestern region of the United States. He said he was attending dog shows in several states and that his unique rare breed had just been accepted in the American Kennel Club and others, and he wanted to be in as many shows as he could to promote the breed. I asked him what the breed was, and he sort of balked a bit and said, "Oh, it's a breed you probably never heard of ... the Nova Scotia Duck Tolling Retriever!" I think I may have screamed like a schoolgirl at a Beatles concert. "No freaking way!" I said, almost crying in excitement. I extolled my admiration for the breed and told him that I had been sort of obsessed with it for a while but that I had never even seen a picture of one. He said, "I would be happy to show my dogs to you."

He went to the motor home, opened the door, and two of the most gorgeous and impressive dogs came running out. They were smaller than I pictured in my head but prettier than I expected. They did indeed look just like very large red foxes. And sweet ... they had such a nice demeanor, especially for hunting dogs which often tend to be high-strung and impatient. One of them had just won a grand champion ribbon and the other a runner-up ribbon. He told me these two dogs were the highest rated and most awarded tolling retrievers in the USA and Canada. I was beside myself with happiness. Not only did I get to see this mythical breed in person, I got to see the two best living examples of it in North America! I could not stop asking myself how this could happen. Those dogs should have been in the cold country of Canada some 3,000 miles away from hot dry West Texas, but here they were right in front of me. It was one of my favorite days ever living in West Texas, and one of the most astounding too. I am pretty sure that if I had not been doing something so intriguing and unusual as growing and selling shrimp in the desert, I may have never seen a Nova Scotia Duck Tolling Retriever in person. I am convinced that when you spend your time doing interesting things, interesting things happen.

We ran the Shrimp Store for nine years. It did a steady business the whole time and kept the bills paid and food in our mouths. It was hard work, though. Don't let anyone tell you otherwise. You don't own the restaurant business, it owns you. The good part is, you do get to go home each day with a satchel of cash, and that is always a great feeling. Despite our best efforts, the shrimp prices across the US continued to decline each year due to the Chinese imports and other foreign production. Every year, there was a new record for the amount of imported shrimp arriving in the US. We reduced the number of acres of shrimp ponds that we ran each year to try and save money and we tried to only grow enough to sell through the wholesale business or the store. This way we got much better margins than by selling through the old routes and using the buyers from the past. We hoped we would control our own destiny this way.

Even though the Shrimp Store had lots of steady business and was popular, it just was not enough to keep the farm producing shrimp in the conventional manner we had been doing before. To add to the misery, the West Texas oil field—especially the Puckett field just south of our close neighbor Fort Stockton—was declining rapidly. Customers were getting fewer, and money was getting tight everywhere in our area. We once again had to think of something to try to slow the seemingly inevitable demise of our unique industry. By now most of the big South Texas shrimp farms and even the giant wild-caught gulf shrimp fleets on the coast like Gulf King out of Aransas Pass were gone. They had all gone broke and bankrupt because of the low prices thanks to imported shrimp. The shrimp game was not the same as it had been, and it was time to think differently about the business. It was time to make a change in how we did things in an effort to construct some kind of long-term future out of all this hard work. I was pretty sure I had an idea of how to do just that.

CHAPTER 11

PERMIAN SEA ORGANICS

T he shrimp store had been very successful, but it was clear that we could not sell enough shrimp to pay the bills for the store and the farm both. West Texas was just not populated enough to move enough shrimp, and the imported foreign shrimp was everywhere and was cheap. The business from the national parks had been a nice addition, but it still cost us a lot to grow shrimp and it was getting more expensive every day. One day we were talking with Rocky, our friend from the Davis Mountains Organic Beef Company. They had been supplying us with ground beef for the last three or four years, which we used for hamburgers in the store. Rocky asked me why we didn't go for an organic certification and sell into the organic markets. Folks with certified products were getting premiums and huge margins if their products were certified organic by the United States Department of Agriculture (USDA).

The idea intrigued me enough that I started studying on the whole process. In 2004, the USDA rules were in place for the certification of plants and crops as well as certification of livestock. There were no separate rules for seafood, so logic would dictate that if it was a marine plant, one could certify it using the plant or crop rules and if it was animal based—e.g., shrimp and fish—then the livestock rules would be the direction to go. Shrimp are actually livestock. Even more challenging was that all livestock had to be certified "100% Organic." Plants, however, could be certified one of two ways: "100% Organic" or just "Organic." "Organic" is the certification if they had some small portion of the process that was not organic, like maybe the seed was not a certified seed but all the fertilizer and cultivation practices were organic. If everything from start to finish was all organic, then you could be certified "100% Organic." Livestock producers did not have a choice. They had to go the 100% route or not at all. We would have to shoot for the USDA's "100% Organic" livestock certification.

Most things that we would be required to do to get a certification would not be difficult. We controlled everything about the shrimp farming—total control regarding all the inputs and outcomes. The West Texas salty groundwater was the cleanest you could find anywhere, and the ground soils had no chemicals from industry or old agriculture, so were also as clean as one could wish. The biggest obstacle for us to deal with would be for us to come up with a 100% organic feed. It would also have to be made by a feed plant that was certified organic. Making a good shrimp feed in the first place is not an easy task. Shrimp have strange dietary requirements, and developing one made with all organic ingredients was going to be the biggest challenge.

We had, in the past, developed a shrimp feed with Cargill that was designed especially for use in West Texas shrimp ponds. It was a very good feed, so I wanted to pattern an organic feed similar to that one. Thankfully, back in my graduate school days in Port Aransas, we made lots and lots of shrimp feeds and fish feeds. It was one of those things that at the time you think, "Why am I having to do this? I will never need this. Feed companies will make the feed and I will just use it. I don't need to know how it was made." Boy, was I wrong about that! I was so glad I made so much feed in that little feed room at the University of Texas Fisheries and Mariculture lab in Port Aransas. The knowledge and experience I gained there would be paramount to getting a feed made. No one in the organic world at the time knew anything about shrimp nutrition, so I had to be the one to figure it out—and figure out where to get all the inputs, which also had to be 100% organic. Then figure out where to find a certified feed mill and how to show them how to make it.

It took a lot of work, but I found everything I needed in one form or another from one place or another. I had to be clever about it. For example: shrimp need cholesterol but cannot make cholesterol like people can, so it is important that it is included in their diets. Cholesterol is what holds cell walls together. Synthetic compounds like synthetic cholesterol are not allowed nor is the use of nonorganic animal products that could possibly provide the cholesterol. I ended up finding an organic cheese that could be my cholesterol supply. It was expired cheese so it could not be sold for human consumption, but it was still good enough to be used as an animal feed product and it was perfect for shrimp feed. We used organic yeast as a protein source as well as organic peanut meal.

While looking for more organic shrimp feed ingredients I had the good fortune to meet a group of folks called the Kansas Organic Producers (KOP) who were a cooperative of farmers who only produced organic crops like soybeans, corn, and grains. This organization was the marketing agent for all the various organic farmers in Kansas and surrounding states. They were like a clearinghouse for organic crops and other organic commodities. Through this group I was able to get most everything I needed, including truckloads of old bread, organic of course, for the carbohydrate component of the feed; the gluten in the bread was a good binder to keep the finished shrimp feed pellet together. I was also informed of a certified organic feed maker in Brownfield, Texas, not too far from Pecos County. They were making feed for organic cattle producers like my friends at Davis Mountains Organic Beef. I went to them and asked if they would be interested in making organic shrimp feed. They seemed interested but indicated that it would be a big challenge to make the shrimp feed since they would have to clean everything in the facility every time they switched from cattle feed to shrimp feed. That was the rule in the organic regulations. The plant must break everything down, clean it all, and set up it again for any new product. As long as I was willing to pay for it, the feed plant was willing to do it. This for sure would make the feed more expensive, but we didn't really have a choice if we were going to make this 100% organic thing a go.

Everything about organic farming was more expensive. The processing plant would also have to take similar measures when processing, but to an even greater degree. Prior to our shrimp being processed (headed and graded and boxed up) the whole shrimp processing plant would have to undergo a cleaning with approved cleaners, then our shrimp could be processed according to the USDA's organic rules. The processing plant was only willing to do this because they wanted someone in the shrimp industry to find some kind of niche that would keep us all afloat. Maybe organic would be it, so we all pitched in to try whatever it took. Once all these steps were in place, which took the better part of a year to complete, and I had presented the organic farm plan and production plan to the third-party certifying agency, we got our certification for USDA 100% organic. We were ecstatic to be the very first aquaculture facility in the United States with an organic certification. Not just the first shrimp farm but also the first aquaculture farm of any kind or species to get this certification!

We were all that remained of the West Texas shrimp farms: Regal Farms and successors, Tucker Farms, and D&T shrimp farms had all gone out of business. We felt alone again, like we did back in the beginning, but we had to keep trying and now we had a new and promising way forward. I had been in this business for more than fifteen years at that point, and I was not going to give up. We changed the name of the business to Permian Sea Organics to reflect the direction we were going. Once word of our certification came out it was a loud buzz in the industry. Everyone had their fingers crossed that this new avenue would bring us the market we needed.

We wanted to start down the organic path the right way, so we signed up for the big, annual organic products show held every year in Austin, Texas. This is where Whole Foods started, and Austin is for sure a place in Texas where you will find the back-to-nature, old school, Mother Earth culture. Most all the buyers and the attendees at the show were indeed this sort of people. To our surprise, though, most of the exhibitors, like us, were just regular, good ole, down-to-earth agriculture folks. Farmers who try to be good stewards of their land and who are kind to their animals. For them it's not some crazy religion, it is just their way of life. I was surprised. We talked to an organic cotton grower, organic pheasant producers, our friends from Davis Mountains Organic Beef, and many others. Seems that most organic producers just wanted to reach this new and burgeoning market and try to get a premium for their products. They all made us very comfortable in this show atmosphere. This was in great contrast to the uncomfortable feeling we got from many of the attendees who were violently organic. Sometimes it almost felt like a political rally rather than a trade show. It was a good thing to have the USDA certified organic label that we could display at this show, because many of the buyers and other attendees were so judgmental about everything that I pitied anyone who didn't have that label. The attendees lectured them on how their products were inferior, and their character was flawed. It was stressful. Overall, it was an interesting show, but we sold nothing and had little interest from the people there other than topical interest in our organic shrimp production. I don't think the organic and natural food world was ready for organic shrimp, but it helped us get ready for doing business with the organic industry and the intense people in it.

While we were busy working on figuring out how to grow organic shrimp, we also had an opportunity to contribute to some conservation efforts that sort of went hand in hand with what we had always done but especially now with the organic farming practices. This project was something near to my heart and that I had always really wanted to do. The desert is a fragile area and there are many species of animals and plants that are in pretty sketchy circumstances in this environment, especially those animals that must rely on a natural water source for their main existence. Natural water is in short supply in this region. The Pecos River pupfish, a little minnow-sized fish, and the puzzle sunflower are two species that are in a pretty bad way around these parts. At the time, these species were on the threatened list and would soon be on the endangered list if something wasn't done to try to increase their populations or get them some sort of haven. Both species required a constant source of brackish water for their critical habitats, and that is something we have always had on our shrimp farm. It only made sense that we could somehow take our man-made environment and provide some kind of refuge for these species. Now, it was not all-out benevolence for sure. There was money available for conservation efforts and habitat construction and we needed all the help we could get. This is the best way to save threatened species: make it profitable or at least cover the costs for the landowners to save the fragile species.

I remember when I was in Florida there was an effort to save the much-endangered Gulf (of Mexico) sturgeon. Government money for saving this fish was very tight, so a clever U.S. Fish and Wildlife Service (USFWS) biologist hatched up a plan to let private fish farmers hatch and grow the sturgeon for profit as long as they released a certain percentage into the wild. There would have been hundreds of thousands of fish released by now if his plan had been accepted, but the plan was held up by environmentalists and government bureaucrats. They did not trust the private sector to help with an endangered fish, so that was the end of any real progress to save that fish. The Gulf sturgeon is still to this day very endangered. I thought government officials misguided and arrogant, as well as shortsighted, about these things. I have great faith in the private sector, so I wanted to work with Fish and Wildlife on these endangered and threatened species and prove their past model for saving species was wrong. I got with the USFWS biologist for my region, and together with Texas Parks and Wildlife personnel we consummated the first public/private deal ever done between USFWS and a private landowner. It was quite an unprecedented landmark agreement. They would provide the money for the project, and I would build and maintain habitats (ponds) for the pupfish. I would also plant areas of puzzle sunflowers along the banks and runoff areas of these pupfish ponds where the critical habitat was similar to the natural areas these plants were found. It would turn out to be a beautiful arrangement and proof that agreements like this can really work.

We built the ponds for the pupfish, and I used our shrimp pond runoff and overflow water to keep these pupfish ponds full. The pupfish thrived and before long there were thousands upon thousands. The sunflowers lived and grew okay, but they were more fickle than the pupfish. Pupfish breed like rabbits—or it really should be rabbits breed like pupfish. See, in the desert a little puddle of water might dry up at any moment due to the arid climate, so the fish that live in that puddle breed like crazy for survival. It's the "eat, drink, (breed), and be merry, for tomorrow we die" scenario. In our stable little man-made pupfish ponds, you could almost walk across it on pupfish, there were so many. They were breeding like, well, pupfish because there were few threats or risk of the water drying up. The only thing that bothered them was the night herons.

Night herons are a type of waterbird, like egrets and the big blue herons that most people see around the water, but unlike these and most other birds, they only come out at night to feed. In addition, in the dark, they make noises similar to dogs barking around the ponds. My longtime hand, Little Manuel, hated night herons. He felt that birds should not sound like dogs, but the reason he did not like them was not because they ate the pupfish in the pupfish ponds. Worse than that, they ate the shrimp in the shrimp ponds too. We would go out sometimes at night and try to scare the herons away. Once Little Manuel smacked one particularly stubborn bird in the head with a piece of PVC pipe. The startled and injured bird made a noise that sounded exactly like someone saying "ow, ow, ow," just like you would say if you smashed your thumb with a hammer. Manuel freaked out and would never again come with me at night to chase away night herons.

Those pupfish loved living on the farm, and I was even presented with honors and a plaque from the U.S. Fish and Wildlife Service for building and maintaining the project. I was proud of that recognition and felt I had done a little favor to some of the weakest of God's creation. We kept this population of fish alive for a couple of years until a scientist that had been doing genetic work on pupfish all over the Texas and New Mexico area concluded that the fish we had in our ponds were not the real Pecos River pupfish. They were not sure which population of pupfish along the Pecos River was the historically correct fish, but they had for sure decided that ours was not the correct one. We were swimming in pupfish, but I guess ours were the redheaded step pupfish variety. We told the biologists that we would just keep the ponds going until they identified the proper pupfish and then we would get rid of these and start over. They never got back to us on that. Meantime, the fish just kept reproducing. The fish flourished so well, in fact, that they ended up getting into some of the shrimp ponds, which led to a situation that really makes me shake my head every time I think of it.

As part of our new organic farm plan, we decided that once the shrimp got to a sellable size, about twenty grams, we would then cast net out seventy-five to a hundred pounds each morning and take them to the Shrimp Store to sell fresh to the customers who wanted fresh shrimp. Unfortunately, in some of the ponds those escaped pupfish would end up getting caught in the nets along with the shrimp, and in numbers sizable enough that it was impractical to pick through the shrimp to get them out. We tried separating them at first, but it just took too long, so we just started dumping them along with the shrimp into the ice and taking the whole mess to the store. When the girls at the store would bag up the shrimp, they would inevitably get a pupfish or two in the bags with the shrimp. Almost no one cared. To most folks it made it seem way more natural and ecological to have some shrimp, seaweed, and maybe even some fish in the bag. We always just told them to pick the fish out and eat the shrimp. All was fine for weeks until one day a local game warden drove up to the store. I knew this guy. He was the game warden out of Monahans, and he always harassed my son and me when we hunted ducks on the Pecos River. He would follow us and then check our guns for lead shot or make sure we had our plug installed in our magazines so that we had only three shells in our guns. He would check our licenses for the federal duck stamp and then he would check our bag to make sure we had not shot too many ducks or the wrong kind. He was a first-class jerk, and the fact he never could find us in violation of any law really pissed him off. He was out for me big time.

He jumped out of his truck, walked fast across the Shrimp Store parking lot, and told me in no uncertain terms that I was in violation of the endangered species act by selling endangered Pecos pupfish. I reminded him that the pupfish was only threatened, not endangered. They would still have been illegal to sell but were only on the threatened list, and we should be accurate, dang it. Then I informed him that these pupfish were frauds and were not the threatened pupfish anyway. None of this deterred him at all. He was very adamant that I was going to prison, and he sure seemed happy about that. Just before he slapped the cuffs on me, I convinced him to call the biologists at the Texas Parks and Wildlife Department and ask them the status of these fish. I worked with these biologists on this project, and I was sure they would be able to set him straight. He resisted at first but eventually he called them, and they indeed set him straight. They told him this particular fish was not the real Pecos pupfish after all, so it would not be a felony for me to have it. They also told him they worked with me, I was a good guy, and he should leave me alone. I know they told him this because I could hear them talking on the phone. The problem was that my character did not make any difference to him. He hated me, and now he was embarrassed in front of his own people so he *really* hated me. He left the Shrimp Store and Imperial, but only for a day. He was not done with me yet, and he would have his revenge.

The next day, back he came—this time with a foolproof plan. He once again drove into the parking lot out front of the store, jumped out of his truck, and asked to see my bait dealer license. After all, in the state of Texas you cannot sell bait without a license issued by his employer, the Texas Parks and Wildlife Department. He just knew he had me now, but I am a pretty good chess player, and I had the checkmate move. I told him that I didn't have a bait dealer license and I didn't need one. I told him that I had a sea-food retail and wholesale license and I was selling shrimp legally. He said, "What about the pupfish? You have to have a bait license to sell those." I laughed and said, "No, sir, I am giving those away. You buy the shrimp; you get free minnows!" BOOM! It blew his mind. I thought his head was going to explode. He told me that he was going to get me—somehow, some way, he was going to get me. What a jerk. I never heard from him again, and word was he retired later that year. I guess I took all the wind out of his sails.

When it was time to start really harvesting the bulk of the crop, we told everyone we knew that we were about to harvest the first-ever organic shrimp. It wasn't long before we started getting calls from all over and soon had several good, high-end retail grocery chains from all over the country ready to purchase our organic shrimp. The first organic crop was beautiful. Not as high of production as before, probably due to our being very conservative. The whole organic venture was new, so we proceeded cautiously, but it was a good crop of shrimp, and they were gorgeous. I wanted to sell all these shrimp to a couple of nice, specialty organic chains that would get us exposure in the retail organic markets and a top price for these one-of-a-kind organic shrimps. I spent some money to get nice boxes and labels made and we put the USDA 100% Organic label right on there for the whole world to see. I didn't need any of this shrimp for the Shrimp Store since I had bought out all of the frozen West Texas shrimp that all the other farms left behind as they were all going out of business. The store was set with plenty of shrimp of all sizes and enough to last till the next shrimp season. These organic shrimp were all going to go for big bucks and hopefully put Permian Sea Organics company in the spotlight again. We needed this to help us start digging out of the financial doldrums that we and everyone else in the industry were going through. The organic shrimp also needed to bring in top dollar because they cost us a fortune to produce under the USDA rules.

I had managed to sell some of the organic shrimp to a high-end grocery chain in El Paso first. This was a nice first sale, but we had a lot more shrimp and needed to bring in a lot more money. I made a deal with an organic grocery chain in California. They said they would buy all of our organic shrimp regardless of size or price.

They wanted them all. It was about 120,000 pounds of tails and at an average of $6.00 per pound it was going to be a nice paycheck. These people were so excited to be able to procure certified organic shrimp for their customers they were just beside themselves.

I had the first loads of these shrimp on two trucks headed out for California when I got a phone call. I was in disbelief. All I could think was how incompetent the government always seems to be and how they can crush the little guy so easily. It turns out that after our certification became known in the public, another shrimp producer in Florida applied for and was about to receive an organic certification too. It seems that some of the well-known natural and organic food stores and companies were caught flat-footed on organic certification of seafood, and they put up some sort of a protest to the USDA. Previously, instead of working with farms and producers in America to get organic seafood on the market, these same companies had apparently invested a lot of money in foreign, organically certified marine products. They knew that this USDA certification would be a huge marketing boost compared to what they were selling, but they were too far invested in the overseas seafood, so their only choice was to go to the USDA and protest that the livestock rules were not detailed enough to include fish and shrimp. They suggested that there should be a separate category of rules and certifications for seafood. They knew this would shut down the USDA certification process and delay any real American organic seafood. It was nothing more than a way to use government bureaucracy to advance the agenda of private business. These were, and are to this day in some cases, very powerful companies with a lot of influence in Washington, DC.

Despite their grassroots, populist, back to nature and the simple life rhetoric, these companies know how to play the game and how to get what they want from the government. They must have had some big pressure on the USDA because soon after their protests, the National Organic Program of the USDA announced that they were going to rescind and retract all organic livestock certifications of seafood until a full set of promulgated rules written strictly for seafood could be worked out, passed through public scrutiny, and then published as official government regulations. This would be at least a three to four-year process at best (turns out to this day there are still no USDA rules for organic seafood). The plan of these big "natural food companies" worked perfectly. This was an instant death sentence for my sales to the organic stores. Before those two trucks full of my organic shrimp ever got close to California, those stores called to cancel the order. They said that their customers would have no confidence in the product now that the USDA rescinded the certifications. It got even worse—the product was actually now illegal because it carried a nonlegal label and selling it as organic could get us a prison sentence. Just owning the shrimp in boxes that were labeled USDA organic was illegal. I can't make this stuff up! I guess the geniuses at USDA didn't think of that at first but in a day or two they made a cleanup announcement that stated we had six months to sell anything labeled organic under our previous certification, but all boxes and labels had to be destroyed after that.

Problem was, it didn't matter anymore. Even though the boxes said organic, and we had been issued a bona fide organic certification, the whole organic retail industry knew that officially the certification had been rescinded and what we had was just conventional shrimp, and conventional shrimp was cheap and widely available. The organic shrimp cost about three times that of conventional shrimp to produce, but now no premium could be obtained for it. I would have to pay to have it all repackaged and sold as regular shrimp at regular shrimp prices. That is, if I could sell any of it at all now that all this had transpired, and the shrimp were part of such a controversy. This would be the final exit wound from the bullet that had entered West Texas shrimp farming on 9/11 five years earlier. There would no longer be any large-scale shrimp farming in the desert again. West Texas desert shrimp farming was in its final dying throes and all the blood, sweat, and tears were just tears now. It brought to mind a quote from the John Chisum character played by John Wayne in the movie *Chisum*. He says prophetically: "There ain't no law west of Dodge City, and there ain't no God west of the Pecos." I was beginning to believe that was true.

A NOTE ON ORGANIC SEAFOOD

As of the writing of this book, the USDA National Organic Program has not yet, more than fifteen years later, developed an organic certification program for seafood—and that's a damn shame. It's embarrassing, actually. In about 2006, they put together an aquaculture working group on which I served for several years to try to make an approved set of rules for organic production. The group got nowhere with it. Every time we agreed on a set of rules and put them out for public comment, there would be a few negative public comments about this rule or that rule. Apparently, the USDA National Organic Program (NOP) didn't want any controversy or disagreement, so any time, which was every time, there were any negative comments from anyone, the NOP officials rejected all the proposed rules. This is strange to me considering there was always contention and disagreement about the other organic certifications within the USDA. There have been disagreements about organic honey and organic yeast, just to name a few. I think they just don't want to be troubled with organic seafood. Or they are being influenced by the big seafood industry. Who knows?

To make the situation even more disappointing, the USDA decided to allow all seafood with an organic certification from anywhere else in the world be accepted for sale and labeled organic in the United States as long as it doesn't say USDA organic. In contrast, all other organic products like meat and vegetables that carry any foreign organic certification must have been certified by rules that comply and duplicate the USDA rules for that product. Because the USDA's NOP rules are the standard for other products, you would think they would simply and rightfully just not allow any organic label on seafood until they have developed and promulgated acceptable standards accepted and trusted by US consumers, but that is not the case. By doing it this way, the National Organic Program has an excuse to say that the lack of US regulation is not standing in the way of the commerce of organic seafood. This somehow gets them off the hook regarding the need for making new rules for organic seafood. More importantly, it gets the USDA off the hook for having to be responsible for oversight and compliance of another commodity they are not interested in. This lack of oversight and initiative by the USDA certainly doesn't give the consumers of seafood products any confidence at all that an organic label on seafood means that it was produced to the standards that American organic consumers want. Unlike on other products, an organic label on seafood in this country doesn't mean a damn thing!

There is not any USDA certified organic seafood and I doubt there ever will be, but I am always proud to educate people that the *only* truly USDA certified organic seafood that has ever been properly certified and properly produced in the United States was produced in West Texas at Permian Sea Organics.

THE END OF SHRIMP FARMING IN WEST TEXAS

With the imported shrimp on the market selling for cheaper prices than we could even grow the organic shrimp, and seemingly no way to find a niche that was profitable, all the shrimp farms eventually had to stop growing shrimp in West Texas. I kept the Shrimp Store going for another four years after the organic debacle, and we grew only a few ponds of shrimp at Permian Sea Organics, just enough to supply what we needed for the store. We stopped selling wholesale shrimp to stores and restaurants and focused on selling everything we produced through the store. I had to lay off Little Manuel after all these years of employment. It was the hardest thing I ever had to do. He had been a faithful, hardworking hand and a good friend to me and Patsy. It felt like I was cutting off my own arm.

All the other farms had already shut down by this time: D&T Farms, Pecos River Aquaculture, and Tucker Farms. All small operations had shut down several years before and the bigger R&G Lighthouse (Regal Farms) shut its doors a few years after they did. Word was that the Arizona farms had all closed and most of the big Texas coastal farms were gone or soon would be. Even the great shrimp boat fleets on the gulf were gone. The only shrimp boats left are Mexican boats. American shrimp farming was all but gone, and it was gone for sure in West Texas. The Permian Sea Shrimp Company was the last surviving farm in West Texas, but we finally had to shut it down for good in 2008. The Shrimp Store had brisk sales up to and including the final year that we had it open, circa 2009. There had been so much promise and so many possibilities. It was heartbreaking to see it all have to go away.

Folks were sad to hear that Permian Sea Organics was gone, but they were devastated when we closed the Shrimp Store. We would have folks come by the house or call and beg Patsy to make them some sautéed shrimp for a party or boil them some shrimp just to satisfy a craving. They were willing to pay any price for their favorite dish or homemade dessert to take home and enjoy once again. It was flattering, but we were done. The restaurant business is a hard business that consumes your life, and trying to make it work when growing the feature item has become cost prohibitive was more than we were willing to endure.

Seven shrimp farms had operated in West Texas: Permian Sea Shrimp Company/Permian Sea Organics, R&G Lighthouse/ Regal Farms, Imperial Shrimp Company, Triton Aquaculture/ Pecos River Aquaculture, the Water District demonstration project, D&T Farms, and Tucker Farms (not counting Arny's one season). I know that some if not most of these farms had profitable years and very nice production levels in many of the seasons prior to the dumping of imported shrimp on the market. During the peak of the production, it certainly looked like there would be an aquaculture industry in West Texas for many years. There was no doubt that the water and land here produced what many believe was the finest, sweetest, best tasting shrimp ever produced either in the wild or on a shrimp farm anywhere. If we could have marketed it on a larger scale and made it a special product in the hearts (and stomachs) of all Texans, then I think it might have survived. Perhaps West Texas shrimp would be the same kind of Texas must-haves as chili, 1015 onions, or brisket BBQ. It all went downhill so fast no one had time or money to get in the marketing game and wave the banner. Cheap shrimp were here to stay, so the only way for West Texas shrimp to stage a comeback would be to define it as a specialty crop with superior attributes that are worth paying more for. I think people would pay more, but when the alternative, foreign shrimp, is so much cheaper, it makes that an even harder sell.

Things may change in the future, and we may feel uncomfortable or unable to source our food from other countries, particularly after the COVID-19 pandemic. It is also possible that the price of shrimp could rise in the future, making West Texas shrimp competitive again. Any of these things could affect the shrimp market and the profitability of farming shrimp. There may come a time when shrimp farming in West Texas gets another look. Maybe under a different set of circumstances and different economic conditions it could have a fighting chance again. I certainly hope so. The shrimp was just too good of a product, and for it to disappear from history is a travesty.

Other aquaculture species also have potential in West Texas. As the shrimp farming was winding down towards the end of the era, I partnered with some of my academic friends to set up a 501c3 non-profit research company called the Organic Aquaculture Institute (OAI). Having a strong belief in the potential for aquaculture and knowing the market advantage of an organic certification, we wanted to continue to investigate other species that might work in this water and this climate. We might have been out-marketed by the Chinese on the shrimp farming, but surely there are other species with potential. We experimented growing some spotted sea trout, tilapia, red drum, and oysters. All species did very well and hold great promise for farming in ponds in this part of Texas. Oysters may be a real promising species now that the micro algae industry has started to mature. The techniques for growing massive amounts of algae, which is what oyster eat almost exclusively, are straightforward and well defined. We were also the first group that learned how to eliminate fishmeal from shrimp feeds. We pioneered organic and sustainable shrimp feeds and published some of that work in scientific journals. The possibilities for the resource are as promising now (if not more so) as they were when I first drove into the Trans-Pecos almost thirty years ago.

If there is any silver lining to the demise of the shrimp farming industry then it is that at least one of the old shrimp farms, my farm, the Permian Sea Shrimp Company farm, has become an algae-producing farm growing a strain of algae that produces omega-3 oils. Traditionally, omega-3s, aka fish oils, come from extracting the oils from fish and refining the oil from the fish—but the fish didn't make the oil, they just concentrate it. In nature, it's algae that make the omega-3 oils in the first place. Algae produce omega-3s and plankton eat the algae, little fish eat the plankton, big fish eat the little fish . . . you get it. It's the food chain. But to get pure, unadulterated, virgin omega-3 oil, you need to squeeze it out of algae.

In 2024, this type of omega-3-producing algae is being grown in the ponds at the old Permian Sea Shrimp Company. The whole farm was repurposed back in 2014 to be an outdoor, open pond algae farm. Not only is the West Texas resource perfect for growing saltwater fish and shrimp, but it is also magical when it comes to growing algae. The University of Maryland did a comparative study using the West Texas water from Permian Sea farm wells and Chesapeake Bay water. They grew algae in both water sources with the same fertilizer and conditions. In the West Texas water, the *Nanochloropsis sp.* strain grew five times faster than the same strain in Chesapeake Bay water. Five times faster! A company called Qualitas Health Inc. grows algae that produces the omega-3 oils for making their "iwi" brand, plant-based omega-3 supplements. After the omega-3 oils are extracted from the algae, the rest of the algae is used for products made from the protein portion of the algae, which is most all of the remaining algae cell.

I have been the Director of Algae Production for iwilife/Qualitas Health Inc. for several years, and I believe that a flourishing algae industry could also become established in the area. Everything that helped the shrimp farming industry will also work for an algae industry, including a cheap, local source of CO_2, which is used as a nutrient in algae farming. Many things can be made from algae, from plastics to food and fuel. Many shoes are now made with foams and urethanes produced from algae. It is a very versatile and interesting industry. Perhaps this industry will catch hold, and West Texas will become the algae capital of the country. It would be a great use of the area's assets and a good diversification for the local economy from the oil industry. West Texas and the Trans-Pecos have a great resource of salty water, electrical infrastructure, good roads, and cheap land. The region is perfect for many types of aquaculture and for growing the very best algae and seafood in the world. The very best shrimp in the world were produced here and could be again under the right circumstances.

This phenomenon was made in Texas and made for Texas, one that everyone should be proud of and cherish and hopefully never forget. It was a brief oasis on an otherwise dry and bleak, unforgiving Chihuahuan Desert landscape. The days of truckloads of fresh, iced shrimp leaving West Texas bound for the processing plants on the coast are gone, as are the hundreds of local pickup trucks and cars each toting an ice chest and waiting in line to buy shrimp fresh out of the ponds on a cool October harvest morning. Gone but hopefully not forgotten are the huge motorcycle and classic car rallies at the Shrimp Store, the tour buses full of elderly adventurers, and the cars parked around the block for fried shrimp Fridays. After more than seventeen years since the last pond harvest of shrimp in Pecos County, one can still meet people in Odessa, Midland, and elsewhere in West Texas, and any mention of the town of Imperial will trigger a response akin to, "Oh, that's where they grow the shrimp," "Can you still get that amazing shrimp down there?" or "The best shrimp I ever ate came from Imperial." People still remember.

Just a few years back a local fellow bought the building where the old Shrimp Store was located, and he converted it into a living space. He made it into his private house, but because so many people, seeing that it was occupied, would stop by and ask about purchasing shrimp and imposing on his privacy, he eventually had to put up a surrounding fence and a sign that says, "Private residence, no shrimp!" The fence and sign are there to this day. The water, the land—it's all still here waiting. And who knows, maybe someday an ambitious fellow with a young family, a solid Texas education, and a head full of dreams might just cruise into town asking a lot of questions.

David from Tucker Farms harvesting West Texas shrimp.

The Shrimp Store.

Patsy serving Bob Phillips of *Texas Country Reporter*.

West Texas Desert Shrimp Growers Association brochure.

Shrimp Store patrons arriving by helicopter.

Patsy and shrimp buyer Dennis at a seafood trade show.

Sea trout and croaker grown by Permian Sea Organics.

Quality Certification Services
(QCS)

Hereby Certifies that

Permian Sea Shrimp Company
P.O. Box 448
Imperial, TX 79743

Meets the strict standards to be

CERTIFIED ORGANIC

Quality Certification Services, a USDA National Organic Program and ISO-65 compliant organic certification
program, has determined, based on a review of the above named entity's application and records, and inspection of
its fields, facilities and processes, that the above named entity meets or exceeds the appropriate and applicable
standards of organic production, handling, and processing. In displaying this certificate, the Certified Entity warrants
that it is in, and will remain in, full compliance with the organic standards set by the
USDA National Organic Program.

Identification Number: 0482L

Effective Date: June 23, 2003

ACREAGE	PRODUCTS
Twelve 4-acre ponds located at 901 E. FM 11 Imperial, TX 79743	LIVESTOCK (See Product Verification for More Information)

PO Box 12311
Gainesville, FL 32604
Phone: 352-377-0133

Certification Coordinator

\\Server\Shared\QCS\Document Binder--QCS\QCS Organic Certificate 04-28-03.p65

Permian Sea Shrimp Companies 100% organic certification.

Algae pond operating on the old Permian Sea Shrimp Farm site.

RECIPES FROM THE SHRIMP STORE

I t may no longer be possible to taste the sweet goodness of a West Texas shrimp or enjoy an evening of dining at the Shrimp Store but we have provided these recipes so you can relive as close as possible the grand days of the Permian Sea Shrimp Company and the Shrimp Store. These are the actual recipes from the store just as we used them, and they are simple to follow. They are not common recipes and that's part of why folks liked them so much. Our menu at the Shrimp Store was unique but very easy and mostly quick to cook. Just like at the store, these recipes are all from scratch. Fear not. By cooking from scratch, you will see how much better the food tastes, and this was the main reason the Shrimp Store was so beloved. This food doesn't taste like everyone else's food. Most all restaurants get their food off the same supply trucks and from the same food service companies, which is why they all taste basically the same wherever you go. We never had any food truck come to our little town. We only cooked from scratch, and it made all the difference. These recipes will make you a shrimp chef in short order. Amaze your friends, win over your enemies, and dazzle your family!

Start with the best shrimp you can find. There is good gulf shrimp if you can purchase it close to the gulf; right off a bay boat is even better because they come in each day, so their shrimp is really fresh. There might be one or two shrimp farms still left on the coast also that produce very good shrimp. The perfect size tails for these recipes is 21–25 tails per pound or 26–30 tails per pound because these were the predominant size we produced in West Texas. I hope these either make a new memory or bring back an old one.

SHRIMP STORE MAIN DISHES

Sautéed Shrimp: The most popular dish from the shrimp store and one of the easiest. Can be made with fresh, head-on shrimp or tails. Do not peel. This recipe is for one to one-and-a-half pounds of shrimp.

In a skillet melt a stick of butter. Add two tablespoons of Cavender's Greek seasoning (or similar), one tablespoon of garlic powder, and one teaspoon of onion powder. Get butter hot but don't burn it. Add shrimp. Turn shrimp every couple of minutes until they are just cooked. They will continue to cook in the hot butter after the fire is off, so shut down the heat just as they are cooked through. Serve on a plate with a good French bread to soak up the butter.

How to eat: First take a shrimp and put the whole thing, shell and all, in your mouth and suck all the garlicky butter off. Then peel the shrimp tail, dip it again in the butter, and eat. It's messy but worth it.

Shrimp Burritos: Use the sautéed shrimp recipe to cook the shrimp but peel the shrimp first. Don't overcook or the shrimp will shrink up small and be tough.

Place about eight of the cooked shrimp in a warm flour tortilla, drizzle some of the garlicky butter on the shrimp, add some sharp cheddar cheese and some salsa. Wrap up and eat. Crazy good burritos.

Boiled Shrimp: Our second most popular dish at the Shrimp Store. Do not use peeled shrimp for this recipe: use shell-on, or heads-on if super fresh.

In a medium-sized pot, fill half full of water. Add one full 12 oz. bottle of Tabasco, ¼ cup cayenne pepper, ½ cup of salt, three teaspoons of lemon juice, ¼ cup garlic powder, and two tablespoons of paprika. Bring to a boil. It should make your nose run, your eyes water, and make you sneeze. The shrimp won't absorb all that spice, but the boil needs to be stout to give good flavor to the shrimp. Add shrimp in small batches of about ½ pound at a time (this will be enough boil for up to five pounds of shrimp). As soon as the shrimp turn pink and float, move them out and put in ice water to cool down immediately. Serve cold, peel, and eat. Be sure to make the red (cocktail) sauce to have with these.

Fried Shrimp: Everyone loves fried shrimp.

Start with peeled shrimp tails. Farm-raised shrimp don't have to be deveined because the farmer takes the shrimp off feed before harvesting, but wild shrimp or Chinese shrimp need to be deveined or you will get sand and grit when you bit into them.

This batter is great for shrimp but also works for fish and oysters as well. In a bowl, add one regular size box of Bisquick, ⅔ cup cracker meal, ¾ of a 5 lb. bag of corn meal, ⅓ cup of onion powder, ¼ cup salt, four tablespoons cayenne pepper, five tablespoons paprika, two tablespoons black pepper, three tablespoons each of onion powder, oregano, and thyme. Mix well then add to it any beer of your choice until it has the thickness of pancake batter. (For frying oysters add some dill pickle juice in place of some of the beer). Mix up a pre dip. This is two tablespoons of corn starch to every cup of flour. Make enough to dredge all your shrimp in before putting in the batter. It makes the batter stick better. Dredge shrimp in the pre dip, then dip into batter until covered and place in hot grease until golden brown. You will eat way too much, I promise.

Shrimp Stuffed and Grilled Poblano Peppers: The most favorite Wednesday special at the Shrimp Store.

For about one pound of peeled/deveined shrimp tails: Take five nice fat poblano peppers and roast the outside and remove the skin. Don't cook the pepper, just burn the skin and peel it off like when you make chili rellenos. Cut open the pepper along the side and remove the seeds. Stuff with about five or so shrimp and shredded cheddar or other cheese. Take three pieces of par cooked bacon (that is bacon that you have partially cooked, but it is still very flexible and pliable and not yet crispy) and wrap each one around the pepper securing with a water-soaked toothpick. Wrap one piece on the top, one in the middle and one at the lower end. Be sure that you seal up the slit in the pepper, so the shrimp doesn't fall out when you cook it. Grill over a gas grill or coals until the bacon is crisp and the pepper is cooked. The shrimp on the inside will be done for sure. The little cages with long handles that are made for cooking peppers on the grill work well for this. Remove the toothpicks and enjoy.

Shrimp Victoria: This is a super easy dish that will impress everyone you serve it to.

Take one pound of peeled shrimp tails. Melt ¼ cup of butter in a skillet and add ½ cup finely chopped onion and your shrimp and sauté for five minutes. Add one cup sliced mushrooms to the sautéed mixture and cook for three minutes more. Mix in one tablespoon flour, ¼ teaspoon salt, dash of cayenne pepper depending on your taste. Then stir in one cup of sour cream (or plain yogurt) and cook gently for ten minutes without boiling. Serve this over rice or pasta. This stuff is amazing.

Shrimp Imperial: Named after the Imperial Mayonnaise used in the dish, not the town. For one pound of peeled shrimp tails.

First make the Imperial Mayonnaise by mixing one cup mayonnaise, ½ teaspoon Tabasco sauce, ½ teaspoon lemon juice, ½ teaspoon Worcestershire sauce, and ½ teaspoon Old Bay

Seasoning. Next make the stuffing by taking one cup of the Imperial Mayonnaise and adding one pound crab meat, ¼ cup breadcrumbs, ¼ teaspoon salt, ¼ cup finely chopped red pepper, and 4 teaspoons finely chopped chives. Split the shrimp (butterfly) so they lay flat on the cooking dish. Lay all the shrimp out in lines in a casserole dish or baking dish and spoon the stuffing onto each shrimp. Brush them with melted butter and white wine and cook at 350° until the stuffing browns. This is a really classy dish and tastes like you're in New Orleans.

Bart's Greek Shrimp: This was a dish I came up with from some similar ideas. The cool thing is you get to sip Jägermeister while you prepare it.

Marinate two-and-a-half pounds of peeled tails in ¼ cup Jägermeister, 1¼ cup dry white wine and ½ cup brandy while you make the other items in the dish. Mince four cloves garlic and ½ cup onion with three or four green onions (green parts also). Sauté in olive oil. When onions are done (opaque) add two diced tomatoes, two teaspoons oregano, a dash of salt and pepper, and the shrimp including the marinade sauce. Cook until shrimp are done and crumble a cup of Feta cheese over the shrimp and reduce heat and cover for five minutes. Serve over seasoned croutons or light pasta. It will change your whole outlook on life . . . but that could be the Jägermeister talking.

Ceviche: This won us Best Cuisine in the Trans-Pecos at the Taste of Texas cooking championship in Alpine, Texas.

Dice up one pound of peeled shrimp tails, place in a bowl and cover with lime juice (fresh-squeezed is best). Let sit in the refrigerator for four hours minimum or up to a day, stirring occasionally to make sure the lime juice gets to all the shrimp. Once shrimp are cooked (they will look pinkish white since the lime juice is cooking them rather than heat) drain the lime juice and then chop and add one cup onion, 1½ cups cilantro, 1½ tomato, one teaspoon salt, ½ cup clamato or V8 juice, and two cups chopped watermelon. Add chopped peppers of your choice and your heat tolerance. We like medium heat, so we use about ½ cup jalapeños. Serve with corn chips, tortillas, in a bowl, on bread or crackers or any way you like. This is addictive. You can prepare ceviche snapper or trout this way too.

CONDIMENTS

Two must-haves for any good shrimp dish are Red Sauce and Tartar Sauce. The store-bought stuff is horrible, so we perfected our own. You can make these in big batches, and they last in the refrigerator a long time and are so good you might even eat them by themselves. Once you perfect these you will never buy the stuff from the store again.

Red Sauce (cocktail sauce):

Take four cups of Heinz ketchup, ½ a jar of Beaver Brand or similar spicy/creamy prepared horseradish. Mix together and taste. Add more horseradish if you like it stout like we do. You can also add a few drops of Tabasco sauce.

Tartar Sauce:

Take one whole, small sweet or mild onion and two cups Best Maid or similar, drained hamburger dill slices (not relish!) and two tablespoons garlic powder. Chop in a food processor until finely chopped. Drain liquid. Add the chopped stuff to four cups of real mayonnaise and a dash of salt to taste. Mix it well and eat 'em up.

SIDE DISHES

We kept the sides simple, so the shrimp was the star of the show. We served various sides through the years, but we always had these two because the customers loved them more than any of the others and they were simple and perfect for serving with most all the shrimp dishes. People would complain if we ran out of these two sides. I mean violently complain!

Shrimp Store Cole Slaw:

Mix two tablespoons crushed dry parsley, ¼ cup of vinegar, 1½ tablespoons Spanish paprika, three tablespoons onion powder, two tablespoons celery salt, ⅓ cup sugar. Add to chopped cabbage or bag of coleslaw greens until thoroughly covered. If making a big batch just double or triple this recipe.

Shrimp Store Potato Salad:

Take three pounds russet potatoes. Boil potatoes until very soft, peel and dice. While potatoes are boiling, finely chop one medium-sized sweet or purple onion, 1½ cups of Best Maid or similar hamburger dills (again, don't use relish), one cup real mayonnaise, ¼ cup mustard, two tablespoons white vinegar, three tablespoons garlic powder, and salt and pepper to taste. Mix all this well with the diced potatoes and refrigerate.

DESSERTS

Patsy's Cheesecake: We served tons of this dessert through the years at the Shrimp Store. Super easy to make.

Take two eggs, one cup sugar, two 8 oz. packages of cream cheese. Soften the cream cheese and mix with other ingredients. Pour into a graham cracker crust. Add chocolate chips to the top or spoon blackberry or strawberry preserves over the top and cut into the cake with a knife. Bake in an oven preheated at 350° for forty to forty-five minutes.

Gooey Butter Cake: Probably the sweetest and richest thing you will ever eat, and eat, and eat.

Crust: one box yellow cake mix, one egg, one stick butter. Mix and press into a baking pan as a crust.

Filling: one pound powdered sugar, two eggs, one 8 oz. package of cream cheese. Mix and pour into crust. Bake at 350° for forty to forty-five minutes. Some folks like the middle pieces and others like the edges with more crust.

Turtle Creek Cheesecake: This is the very best no-bake cheesecake you will ever put in your mouth and that is no exaggeration! Makes three cakes, and you will be happy that it does.

For the cakes: Mix one pound (16 oz.) cream cheese, half a pound of powdered sugar, and one tub of Cool Whip (16 oz.). Pour into a graham cracker crust. Brown one cup of coconut and one cup chopped pecans in a skillet. Mix these with one small jar of warmed caramel and then spoon the mixture on top of the cheesecakes.

Hershey Bar Pie: This was from my childhood and what I often asked for instead of a birthday cake.

In a pan or skillet on the stove top melt (low to med heat) twelve Hershey milk chocolate bars with almonds, twenty-seven large marshmallows, and ½ cup milk until it's all melted together and blended. Allow to cool to a lukewarm temperature. Whip up one cup of cream and then fold into the melted chocolate, mixing well. Pour into a graham cracker pie crust and refrigerate.

ACKNOWLEDGMENTS

I would like to first thank Dr. Travis Synder of Texas Tech University Press who enthusiastically helped me develop this story and save it for posterity. It started out as a bunch of fond (and not so fond) memories, and he encouraged me to turn it into a real memoir of the period. I am thankful to all the folks at Texas Tech University Press who helped make this possible, especially Carly Kahl and Christie Perlmutter, who spent a good deal of time fixing my mistakes and making this readable.

I would like to thank Dina López, Fritz Jaenike, and any others who read early copies of the manuscript and gave good suggestions to make it better. Thanks also to Rachel Lyon for indexing on short notice and with a tight timeline.

I especially want to thank my professors and mentors from my days at Texas A&M–Corpus Christi and the University of Texas Marine Science Institute: the late Dr. Connie Arnold, the late Dr. Wes Tunnel, Dr. Joan Holt, and especially Dr. David McKee who through his writing and publishing inspired me to take up the pen and put this to paper. It seemed the least I could do to honor the ones who gave me the foundation to have a successful career in marine aquaculture.

Lastly, I want to thank Dr. Miles Palmer, Mr. Doug Lynn, and Dr. Isaac Berzin who came along after the smoke cleared and agreed with me that the future for West Texas aquaculture looked like green algae, which has allowed me to continue the journey in the Trans-Pecos.

APPENDIX: RECIPES
(SHORT FORM)

Sautéed Shrimp
8 oz. (1 stick) butter
2 T. Greek seasoning
1 T. garlic powder
1 t. onion powder

Melt butter in skillet. Add Greek seasoning, garlic powder, and onion powder. Heat but don't burn butter. Add shrimp. Turn shrimp every couple of minutes until they are just cooked. Serve with bread.

Shrimp Burritos
8 sautéed shrimp (see recipe above)
Flour tortillas
Shredded cheddar cheese
Salsa

Place cooked shrimp in warm tortilla, drizzling on garlic butter. Add cheese and salsa and wrap.

Boiled Shrimp
5 lbs. shrimp (tails-on or very fresh heads-on shrimp only, not pre-peeled)

12 oz. Tabasco sauce
¼ cup cayenne pepper
½ cup salt
1 T. lemon juice
¼ cup garlic powder
2 T. paprika

Fill medium-sized pot half full of water. Add Tabasco sauce, salt, lemon juice, garlic powder, and paprika. Bring to a boil. Add shrimp in small batches of about ½ pound at a time When shrimp turn pink and float, put in ice water to cool down. Serve cold with cocktail sauce.

Fried Shrimp
Peeled and deveined shrimp tails
Dip
1 box (40 oz.) Bisquick
⅔ cup cracker meal
¾ of a 5 lb. bag of corn meal
⅓ cup onion powder
¼ cup salt
4 T. cayenne pepper
5 T. paprika
2 T. black pepper
3 T. onion powder
3 T. oregano
3 T. thyme
Beer to consistency
Oil for frying
Pre-dip
2 T. corn starch per 1 cup flour

In large bowl, place Bisquick, cracker meal, corn meal, onion powder, salt, cayenne pepper, paprika, black pepper, onion powder, oregano, and thyme. Mix well and add beer to the consistency of pancake batter. (For frying oysters, add some dill pickle juice in place of some of the beer.) Dredge shrimp in pre-dip. Then dredge shrimp in batter until covered. Place in hot oil and fry until golden brown.

Shrimp Stuffed and Grilled Poblano Peppers
1 lb. peeled/deveined shrimp tails
5 large poblano peppers
Shredded cheddar (or other) cheese
15 slices partially cooked bacon

Roast peppers and remove skin. Cut open peppers along the side and remove seeds. Stuff with about five shrimp each and cheese to taste. Wrap three pieces of bacon around each pepper, securing with a water-soaked toothpick. Grill over a gas grill or coals until the bacon is crisp and the pepper is cooked.

Shrimp Victoria
1 lb. peeled shrimp tails
¼ cup butter
½ cup finely chopped onion
1 cup sliced mushrooms
1 T. flour
¼ t. salt
Dash cayenne pepper
1 cup sour cream (or plain yogurt)

Melt butter in skillet. Add onion and shrimp; sauté for five minutes. Add mushrooms and cook for three minutes more. Mix in flour, salt, cayenne pepper to taste. Stir in sour cream and cook gently for ten minutes without boiling. Serve over rice or pasta.

Shrimp Imperial

2 pounds shrimp, peeled and deveined

Butter

White wine

<u>Imperial mayonnaise</u>

1 cup real mayonnaise

½ teaspoon Tabasco sauce

½ teaspoon lemon juice

½ teaspoon Worcestershire sauce

½ teaspoon Old Bay seasoning

Mix all ingredients above.

<u>Stuffing</u>

1 cup of Imperial mayonnaise (see above)

1 lb. crab meat

¼ cup breadcrumbs

¼ t. salt

¼ cup finely chopped red pepper

4 t. finely chopped chives

Mix Imperial mayonnaise with crab meat, breadcrumbs, salt, red pepper, and chives. Split the shrimp (butterfly) so they lay flat on a cooking dish. Spoon stuffing onto each shrimp. Brush with melted butter and white wine and cook at 350° until the stuffing browns.

Bart's Greek Shrimp

2 ½ pounds peeled shrimp tails

¼ cup Jägermeister

1¼ cup dry white wine

½ cup brandy

4 cloves garlic
½ cup onion
3 or 4 green onions (including green parts)
2 diced tomatoes
2 t. oregano
Dash each salt and pepper
1 cup Feta cheese
Olive oil

Marinate shrimp in Jägermeister, wine, and brandy. Mince garlic and onion with green onions. Sauté in olive oil. When onions are done (opaque) add tomatoes, oregano, salt and pepper, and shrimp including marinade. Cook until shrimp are done. Crumble Feta cheese over shrimp, cover, reduce heat, and cook for five minutes. Serve over seasoned croutons or light pasta.

Ceviche
1 lb. diced peeled shrimp tails
Lime juice (preferably fresh squeezed)
1 cup chopped onion
1½ cups cilantro
1½ cups tomato
1 t. salt
½ cup Clamato or V8 juice
two cups chopped watermelon
½ cup chopped jalapeño (or other) peppers

Place shrimp tails in bowl and cover with lime juice. Let sit in refrigerator for four to twenty-four hours, stirring occasionally. Cook shrimp and drain lime juice. Add onion, cilantro, tomato, salt, Clamato juice, and watermelon. Add chopped peppers to taste. Serve with corn chips, tortillas, or crackers.

Red Sauce
4 cups ketchup
½ jar Beaver Brand or similar spicy/creamy prepared horseradish

Mix and add horseradish or a few drops of Tabasco sauce to taste.

Tartar Sauce
1 whole small sweet or mild onion
2 cups Best Maid or similar drained hamburger dill slices (not relish)
2 T. garlic powder
4 cups mayonnaise
Dash salt

Chop in food processor until fine. Drain liquid. Add mayonnaise and salt to taste. Mix well.

Shrimp Store Coleslaw
2 T. crushed dry parsley
¼ cup vinegar
1½ T. Spanish paprika
3 T. onion powder
2 T. celery salt
⅓ cup sugar
Chopped cabbage or bag of coleslaw

Mix parsley, vinegar, paprika, onion powder, celery salt, sugar. Add to cabbage or coleslaw greens until thoroughly covered

Shrimp Store Potato Salad
3 lbs. russet potatoes
1 medium sweet or purple onion
1½ cups Best Maid or similar hamburger dill pickle slices (not relish)

1 cup mayonnaise
¼ cup mustard
2 T. white vinegar
3 T. garlic powder
Dash salt and pepper

Boil potatoes until very soft; peel and dice. While potatoes are boiling, finely chop onion and hamburger dills. Add mayonnaise, mustard, vinegar, garlic powder, and salt and pepper to taste. Mix well with diced potatoes and refrigerate.

Patsy's Cheesecake
2 eggs
1 cup sugar
2 8 oz. packages cream cheese
Prepared graham cracker crust
Chocolate chips or preserves (strawberry or blackberry) for topping

Soften cream cheese and mix with eggs and sugar. Pour into graham cracker crust. Sprinkle or spoon chocolate chips or berry preserves over the top and cut through the filling with a knife. Bake in pre-heated oven at 350° for forty to forty-five minutes.

Gooey Butter Cake
Crust
1 box yellow cake mix
1 egg
8 oz. (1 stick) butter
Mix and press into a baking pan.
Filling
1 lb. powdered sugar
2 eggs

1 8 oz. package cream cheese

Mix and pour into crust. Bake in preheated oven at 350° for forty to forty-five minutes.

Turtle Creek Cheesecake
Note: Makes three cheesecakes.
1 lb. (2 packages) cream cheese
1 cup powdered sugar
16 oz. Cool Whip
3 prepared graham cracker crusts
1 cup grated coconut
1 cup chopped pecans
1 small jar caramel sauce

Mix cream cheese, powdered sugar, and Cool Whip. Pour evenly divided into graham cracker crusts. Brown coconut and pecans in skillet. Mix coconut and pecans with warmed caramel and drizzle mixture on top of cheesecakes.

Hershey Bar Pie
12 chocolate bars with almonds (1.45 oz. each)
27 large marshmallows
1 cup whipping cream
Prepared graham cracker crust

In a pan or skillet melt (low to med heat) chocolate bars, marsh-mallows, and milk until blended. Allow to cool to lukewarm temperature. Whip cream and fold into melted chocolate, mixing well. Pour into graham cracker crust and refrigerate.

INDEX

ABOUT THE AUTHOR

Bart Reid is a marine biologist and entrepreneur with more than forty years of experience in all types of aquaculture, ranging from shrimp to fish to algae. He holds BS and MS degrees from Texas A&M University. In the 1980s he helped develop the hatchery techniques for breeding redfish and speckled trout that are used by the state hatcheries to restock the bays and waterways in the Gulf Coast region. After many years in the shrimp farming business, he is now involved in algae culture for the production of both nutraceuticals and bioplastics. He is also the president and CEO of RRMercator, selling Bart's Bay Armor protective wading boots for bay fishing. Bart has been awarded two patents: one for macro algae production and one for puncture-resistant materials.

He is an avid hunter and fisherman and divides his time between the Trans-Pecos area and Port Mansfield, Texas, on the coast where he has a house and boats.

Printed in the USA
CPSIA information can be obtained
at www.ICGtesting.com
LVHW051919170924
791294LV00003B/434